Differentiating

in Measurement

Differentiating

in Measurement

A **Content Companion** for

- Ongoing Assessment
- Grouping Students
- Targeting Instruction
- Adjusting Levels of Cognitive Demand

Jennifer Taylor-Cox

HEINEMANN
Portsmouth, NH

Heinemann
361 Hanover Street
Portsmouth, NH 03801–3912
www.heinemann.com

Offices and agents throughout the world

Library of Congress Cataloging-in-Publication Data
Taylor-Cox, Jennifer.
    Differentiating in measurement, preK–grade 2 : a content companion for ongoing assessment, grouping students, targeting instruction, and adjusting levels of cognitive demand / Jennifer Taylor-Cox.
        p. cm. — (Differentiating in number & operations and the other math content standards, preK–grade 2)
    Includes bibliographical references and index.
    ISBN-13: 978-0-325-02187-4
    ISBN-10: 0-325-02187-2
    1. Mathematics—Study and teaching (Preschool).   2. Mathematics—Study and teaching (Primary).   3. Measurement—Study and teaching (Preschool).   4. Measurement—Study and teaching (Primary).   5. Individualized instruction.   I. Title.
QA135.6.T3495 2008
372.7—dc22   2008017188

Editor: Emily Michie Birch
Production: Lynne Costa
Cover design: Jenny Jensen Greenleaf
Typesetter: Publishers' Design & Production Services, Inc.
Manufacturing: Louise Richardson

Printed in the United States of America on acid-free paper

12   11   10   09   08   VP   1   2   3   4   5

CONTENTS

CHAPTER 1

How to Use This Book

The Purpose of the *Measurement Content Companion*

Why is there a Content Companion?

The purpose of this *Measurement Content Companion* is to provide prekindergarten through second-grade mathematics educators with the tools needed to target instruction through differentiation in measurement. The main book (*Number & Operations*) offers ideas, techniques, and strategies to use when targeting instruction in number and operations. The *Measurement Content Companion* allows educators to understand how to use these same ideas, techniques, and strategies when teaching measurement. The *Measurement Content Companion* offers specific content examination, student work samples, and lesson ideas. The rationale behind the *Content Companions* is that while number and operations is a critically important math standard, it is not the only standard that prekindergarten through second grade math educators teach. Algebra, geometry, measurement, and data and probability are also very important content standards.

How to Use This *Measurement Content Companion*

How do I use this book?

The easiest way to use this book is to follow the number and operations book, looking for the cross-reference boxes in the margins that indicate where other mathematics content explanations and examples are offered in the content companion. Each time the reader sees this cross-reference box in the number and operations book, the reader can turn to the corresponding page in this *Content Companion* to find specific information pertaining to measurement. There are far too many books out there that attempt to explain to readers that similar strategies can be applied in all areas of mathematics, but these resources often fail to provide specific examples. Likewise, some resources attempt to cram all content into one resource, but these resources are often limited and incomplete. To address these problems, the *Content Companions* provide a comprehensive set of resources that prekindergarten through second grade educators can use to target instruction through differentiation in mathematics.

The chapters in the *Measurement Content Companion* align with the chapters in the number and operations book. Chapter 1 offers an overview. Chapter 2 provides specific content information and examination pertaining to measurement. Chapter 3 directly addresses targeted instruction using informal preassessment. Chapter 4 involves grouping students. Chapter 5 tackles levels of cognitive demand. Chapter 6 focuses on learning frameworks. Chapter 7 provides personal assessment examples and situations.

While any one of the *Content Companions* could be used as a stand-alone resource, it is best understood within the context of the number and operations book. Educators are encouraged to use the number and operations book along with the four *Content Companions*—algebra, geometry, measurement, and data analysis and probability—to learn how to target instruction through differentiation in mathematics for prekindergarten through second-grade students.

CHAPTER 2

Measurement in Prekindergarten Through Second Grade

The Importance of Measurement

Measuring up

Measurement is math in action. We use measurement every day of our lives—"Measurement is one of the most widely used applications of mathematics" (National Council of Teachers of Mathematics 2000, 103). Determining size, amount, and degree requires the real-life application of number sense, counting, operations, and geometry. When young children measure, they need to use hands-on, minds-on approaches. They need to connect the mathematics in ways that make sense. Understanding the measurable attributes of objects requires various and diverse experiences. Children need to experience measurement by using appropriate tools and applying accurate and efficient techniques and formulas.

Understanding the Measurable Attributes of Objects and Using Measurement Tools

There are many measurable attributes of objects. We can measure length, weight, capacity and volume, area and perimeter, angle, temperature, and time. Each of these branches of measurement gives us information about the size, amount, and degree of an object. The tools we use to obtain these measurements vary in type and precision. Our understanding and communication of measurement manifests itself through comparison, order, and units. Let's take an in-depth look at each branch of measurement and the associated measurement tools, units, and levels of understanding.

Length

Length is the measurement of something from end to end. Width, it follows, is the measurement of something from side to side. While recognizing that length typically refers to the longest part or greater dimension of an object, we need not overemphasize the distinction between length and width. Instead we should focus on using these dimensions to determine the measurement. Length and width are attributes of two-dimensional objects. If we are measuring three-dimensional objects we also need to consider the height and depth of the object. If we are working with one-dimensional figures, we need only focus on length. Even though length is one of the dimensions, it is also the broad category label for all of these dimensions (length, width, and height).

The earliest stages of measuring length involve comparison. Which shoe is longer? Which box is taller? Which table is widest? These comparison words involve adding a suffix to the adjective to form comparative degrees—*er* (as in *longer*) and *est* (as in the superlative degree, *longest*). The comparative endings only make sense in relation to something else. If we say *the bookshelf is longer*, it doesn't make sense because we are left with the question *longer than what?* But if we complete the idea by saying *the bookshelf is longer than the rug*, we have a clearer picture of what *longer* means in this context. Helping children learn how to complete their measurement comparisons is beneficial. We need to model this idea and use questioning to help students come to understand the importance of comprehensive comparative statements in measurement.

Young children often see vertical length (height) more distinctly than they see horizontal length. If the teacher wants the children to compare two towers of cubes, the difference is often more obvious to the children when the towers are presented vertically. Perhaps this is just a matter of the children's perspective, or perhaps it has to do with clarity and familiarity in the baseline. In any event, if children are having difficulty noticing differences in length, try presenting the comparison objects in a vertical fashion.

Substantial differences in length, width, or height are often more apparent to young children than are slight differences. In the comparison of two towers of cubes, the children are more likely to notice bigger differences in size. For example, a tower of ten cubes compared to a tower of three cubes is more obvious than a tower of five cubes compared to a tower of four cubes. However, if the objects are meaningful to the children, they will notice even the smallest differences. Say the teacher gives the tower of five cubes to one child and the tower of four cubes to another child. Most likely there will be protests— "Hey, her tower is taller than mine" or "Why is my tower shorter?" The children notice slight differences in length, width, and height when the differences affect them personally.

Comparing length also involves understanding when two objects are the same length. One way to verify that objects are equal in length is to compare them side by side. If the objects are moveable, the comparison is less complicated than if the objects cannot be moved. In situations where the objects are not mobile, the length can still be compared using other objects (e.g., string, blocks, and links).

The next level of understanding how to measure length involves ordering. Once children can compare the size of two objects, it is time to start using more than two objects. When faced with three or more objects, the comparison moves up a notch in complexity. There are more comparisons that need to be made and this information needs to be used to place the objects in ascending or descending order. The towers are ordered from shortest to tallest. The name tags are ordered from longest to shortest. It does not matter if you begin with the smallest or the largest, what matters is that the objects are placed in meaningful order. Typically the greater the number of objects that need to be ordered, the more complex the task is. Slight differences (or no differences) in the lengths of the objects (rather than considerable differences) also increase the difficulty of the task.

After working with comparing and ordering length, children can begin to work with nonstandard units. Nonstandard units are any objects that can be used to communicate size, amount, or degree. These objects serve as the units and as the tools for measuring with nonstandard units.

To measure a pencil, we can use various nonstandard units (see Figure 2–1).

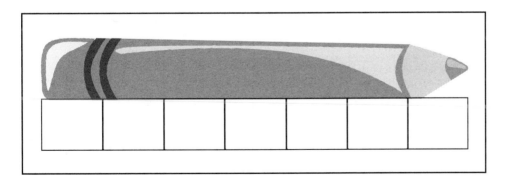

Figure 2–1
*Measuring a Pencil
with Nonstandard
Units*

The pencil is seven tiles long. Even when using nonstandard units, the use of the unit is significant. The pencil is not seven long; it is seven tiles long. Sometimes children try to use different types of nonstandard units at the same time. The problem is apparent because the units are not the same size (see Figure 2–2). *The pencil is nine* does not give us enough information. While you can accurately report that these specific items are the same length as the pencil, it is convoluted and time consuming to communicate that information. Helping children learn why it is more efficient to use the same objects as the nonstandard units is part of the process of learning how to measure length.

Accuracy is enhanced when the nonstandard units are positioned in a uniform manner. When spacing is not regular, the measurement is not accurate (as in Figure 2–3).

Even though there are eight blocks next to the pencil, the pencil is not eight blocks long. The intervals are not consistent, causing the measurement to be fictitious. Additionally, sometimes children do not line the objects up from end to end, but place blocks starting in the middle. Sometimes they continue placing blocks all the way around the object (finding the perimeter rather than the length).

It is important to recognize that all of the mistakes that children make when measuring length are important steps in the learning process. As teachers, our job is not to take over and do the measuring for the students. Instead

Figure 2–2

Attempting to Measure a Pencil with Different Sizes of Nonstandard Units

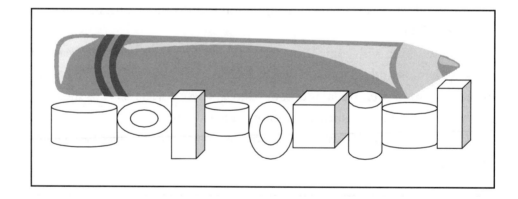

Figure 2–3

Attempting to Measure a Pencil with Differently Spaced Nonstandard Units

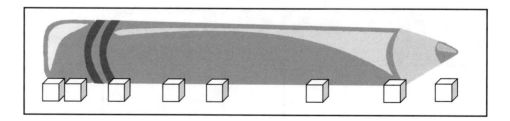

we take each error as a learning opportunity. We ask questions. We invite the students to share and compare ideas, procedures, and results. We encourage them to think mathematically as they learn to measure length.

During this work with nonstandard units, it is important for children to come to understand that the size of the unit is related to the quantity of the unit. The bigger the unit, the less you will need to measure an object. Likewise the smaller the unit, the more you will need to measure the same object. This critical aspect of measurement is not a rule that we should have children memorize. Instead, it is a level of understanding that children reach given the right experiences and the right amount of guidance and facilitation during those experiences.

In addition to using objects for nonstandard units, children can use body measurements. It is interesting to share with children that throughout history people have used themselves to measure length. They used finger widths, palm widths, hand spans (from thumb to little finger), cubits (from elbow to fingertip), and fathoms (from fingertip to fingertip with outstretched arms) to measure length. They even used inch (thumb width), foot (from heel to toe), and yard (from nose to fingertip when arm is outstretched), which became the standard units that many of us use today.

There are many tools that help children create a bridge from measuring with nonstandard to standard units. One of my favorite bridging tools is base-ten blocks (originally designed by Zoltan Dienes; Dienes 1960; Dienes and Goldring 1971). The base-ten blocks are fabulously designed in that the ones units are centimeter cubes. Ten centimeter cubes is the same length as one ten rod, which is the same length as one decimeter. Children can use these manipulatives in the same way that they used nonstandard units. They just happen to be standard units of measure. Likewise, one-inch tiles offer nonstandard means of measuring with standard units. There are also some measurement manipulatives that connect together. The process of attaching the units together to create a measurement tool is powerful. The idea of bridging from nonstandard to standard units by using standard-sized manipulatives is an excellent way to help children be successful.

Common tools for standard units used for measuring length include rulers, meter sticks, yardsticks, tape measures, trundle wheels, and odometers. These tools can either help children understand standard units of measure or add to children's confusion.

The problem with rulers is that there is typically so much information on them that they are difficult to use. What we need are rulers that just show whole units and separate rulers that show fractional parts of the units. I have seen a few rulers that attempt to have parts of the ruler connected like pages in a book, but these are just too cumbersome for young children to use. The simpler the ruler, the easier it is for young children to use.

For some children, learning how to use a ruler is difficult. They are not sure where to start the measurement. They do not know how to line up the ruler because they do not yet understand enough about the tool. One of the ways to deal with this problem is to encourage children to construct their first rulers out of manipulatives or grid paper. Attaching or lining up the units to create a ruler (which also happens to be a number line) helps young children understand the purpose of the tool, and thereby opens the door for understanding how to use the tool.

A trundle wheel is a fantastic tool for measuring substantial length. It is like a meter stick or a yardstick that is circular. As you walk, you push the wheel to measure distance. The wheel clicks after every rotation. Some trundle wheels are metric and others are customary. The wheels are usually graduated into the respective smaller measurements. A mechanical odometer functions in a similar way. It is a small geared mechanism with a gear ratio that measures length in terms of miles or kilometers.

Deciding which tool is best for measuring length also involves knowing which unit is optimal for the situation. What if we need to measure a desk, the classroom, the whole school, and the entire community? Inviting children to think about which units are small enough or large enough to measure the length of any given object encourages them to apply problem solving and reasoning to the measurement situation.

Weight

The term *weight* actually has three different meanings: *gravitational force*, *mass*, and a *calibrated mass*. In physical science we understand that the weight of something is the gravitational force acting on it (which is measured in newtons). In our daily experiences with economics (specifically the trading of goods) weight means the measure of the amount of material in an object, which is the object's mass. In mathematics we often use weights (objects that weigh a specific amount and are usually made of metal) to measure the weight of an object. These definitions add clarity to the concept of weight and allow us to use the term *weight* when referring to mass, force, or a standard metallic solid.

Children learn about weight by pushing, pulling, lifting, holding, and carrying objects in their environments. The beginning stages of measuring weight involve comparison. Which shoe is heavier? Which book is lighter? Which backpack is heaviest?

These early comparisons of weight usually involve holding one object in each hand. If the difference in weight is substantial enough, we can tell which object is heavier and which object is lighter. In this way, the person serves as the tool for measuring weight. As the differences in weight become less obvious, we need to turn to other tools that will help discern the weight difference. A simple balance or a pan balance often functions quite well for these early experiences with weight (see Figures 2–4 and 2–5).

These balances help children understand weight comparisons because the differences in weight are apparent in the position of the arm of the balance. When comparing a heavy object to a light object, one side of the balance goes down and the other goes up. The arm of the balance is diagonal. As one kindergartener shared with me, "It is just like a seesaw. The heavy one goes down." When the objects weigh the same amount, the arm of the balance is horizontal, showing equality.

The bucket balance is designed to hold greater amounts, as is shown in Figure 2–6. Although the bucket balance allows for greater amounts, the problem is that the differences in weight are not as obvious to young children as they are with the simple balance. Because of the design, there is little room for

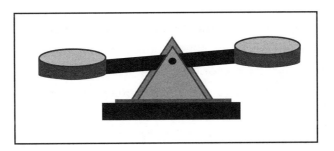

Figure 2–4
An Example of a Simple Balance

Figure 2–5
An Example of a Pan Balance

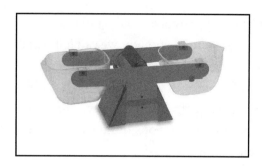

Figure 2–6 *An Example of a Bucket Balance*

movement. Even if one bucket contains something that is *much* heavier than the items the other bucket contains, the arm of the balance is only slightly diagonal. Essentially, weight differences are not as pronounced as they are when using the simple or pan balances.

The next level of understanding how to measure weight involves ordering. Just like working with length, once children can compare the weight of two objects, it is time to move onto using more objects. Ordering objects by weight is a multifaceted task. Each object is commonly weighed against other objects *separately* to learn the comparative weights. Each use of the balance becomes a different step in the process of ordering the objects by weight. If more objects need to be ordered or if the differences in weight are minimal rather than profound, the multifaceted task of ordering by weight is further compounded.

After working with comparing and ordering weight, children begin to work with nonstandard units. Nonstandard units are any objects that can be used to communicate size, amount, or degree. These objects serve as the units and as the tools for measuring with nonstandard units. Children place the objects on balances to find which items weigh more and which items weigh less.

As with other measurement tools, children begin by exploring with the balances. They often keep adding more and more objects, watching the changes in the balance. It may seem like they do not know what they are doing because they just keep putting more and more items on the balance. However, this exploration and experimentation with cause and effect helps them to better understand the functions of the balance. They learn how the position of the item in the pan or bucket can affect the weight. They learn that if the units are lighter, they will need more of them than if the units are heavier.

Using the unit to report the weight is a critical aspect of using nonstandard units. Instead of *the pencil weighs four*, students need to report that *the pencils weighs four cubes*. Using the nonstandard units to report the weight serves as a bridge to using standard units. Using the weights that are calibrated into standard measurements, students can learn that *the pencil weighs seven grams*. Each gram is a separate weight, which makes it easier for the students to connect standard units to nonstandard units. The task becomes more difficult when any one of the weights represents more than one unit (e.g., a single weight that equals ten grams).

Weight with standard units becomes even more complex when we move from using balances to using scales. Scales can be analog or digital. Just like clocks, these platform scales show the same measurement in two different formats (see Figure 2–7).

The analog scale has an arrow that shows the weight on a dial and the digital scale shows the digits to indicate the weight. It is often easier for children to *read* the digital scale, but this ease does not necessarily mean they actually understand it. The analog scale helps children use their number sense to understand the measurement because they can view it as a circular number line. They can see the weight in context and in relation to other amounts on the dial.

Personal scales also come in analog and digital varieties, as shown in Figure 2–8. Like the platform scales, these scales indicate weight with an arrow on a dial or with digits. These personal scales can show substantial weight.

Figure 2–7
Some Examples of Analog and Digital Platform Scales

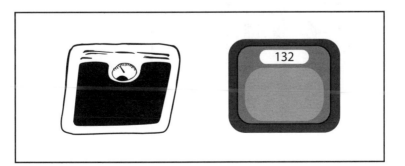

Figure 2–8
Analog and Digital Personal Scales

Spring scales are another tool that can be used to measure weight (see Figure 2–9). Hanging the object on the hook causes the spring to show the weight on the number line. The function of the spring scale highlights the relationship between force and stretch. The weight of the object is compared against the force of a standard spring. The spring scale allows us to measure weight using force and stretch.

Deciding which tool is best for measuring weight also involves knowing which unit is optimal for the situation. What if we need to measure the weight of a bug, mouse, chicken, dog, and an elephant? Inviting children to think about which units are small enough or large enough to measure any given object encourages them to apply problem solving and reasoning to the measurement situation.

Capacity and Volume

Capacity and volume are interrelated, but somewhat distinct concepts. Volume refers to the space an object takes up; capacity refers to the amount of pourable substance a container can hold. Usually we talk about the capacity of a container and the volume of a three-dimensional object that occupies a specific amount of space. Even though many people use the two words interchangeably, we help children understand the concepts more clearly when we identify and describe the specific type of measurement. For young children, the focus on capacity builds foundations for understanding volume.

Comparing the capacity of two containers is the beginning stage. Children need to decide which container holds the most, the least, or if the capacities are equal. The pourable substance is often water, but sand, rice, beans, and unpopped popcorn also serve as great pourable substances to use in comparing the capacity of containers. When we give two containers to children asking

Figure 2–9
Spring Scales

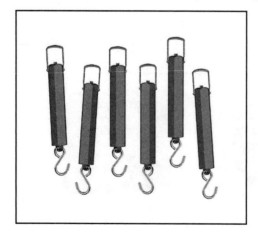

which container holds the most, they often decide based upon the appearance of the containers. Younger children sometimes think tall containers hold more than short containers, even when the shorter container is very wide and deep. The focus on height as *bigger* or *having a greater capacity* is connected to the Piagetian concepts of conservation and centration (Piaget 1977, 1983). When children have the ability to conserve, they understand that a tall, thin, twelve-ounce container holds the same amount as a short, wide, twelve-ounce container. When children think the tall container holds more than the shorter container it is because of centration—they are focusing on one characteristic (height).

To help children learn that shorter containers can have equal or greater capacity than taller containers, we need to teach children to *test* the capacity by filling containers. Adorably, many children will fill both containers simultaneously and announce that the taller container holds more. Of course, this way of testing the containers' capacities does not prove which container holds more than the other container. To truly test the capacity of the containers, children need to fill one container and pour the substance into the other container so that a comparison can be made. Learning how to test containers' capacities does not guarantee knowledge of which container holds the most or least because the child may still concentrate on the height of the containers rather than comparing the substance in each container; however, teaching children how to test the capacity of containers is an important step. The more experiences children have with pouring, observing, discussing, and reasoning, the further they will advance in their understanding of comparing capacity.

When comparing the capacity of containers, children should use words and ideas such as *empty* and *full*. They can extend these ideas by describing the capacity of specific containers as *almost full* or *almost empty*. Sometimes young children say a container is *a little bit full* when they actually mean that the container only has a small amount in it (or is *almost empty*). Providing clarity through our modeling of descriptions and our questioning helps children come to understand capacity.

Ordering containers by capacity involves comparing more than two containers. For example, the buckets are ordered from most shallow to the deepest. Ordering containers by capacity is a multifaceted, complex task. Typically, children learn to make predictions in the order and then fill and pour to check the predictions. Each time they fill a container and pour the substance into another container is a different step in the process of ordering the containers by capacity. If more containers need to be ordered or if the differences in capacity are minimal rather than profound, the multifaceted task of ordering by capacity is further compounded.

After working with comparing and ordering capacity, children can begin to work with nonstandard units. Nonstandard units are any objects (interpreted as pourable substance) that can be used to communicate size, amount, or degree. These objects serve as the units and tools for measuring with nonstandard units. Children place the objects in the containers to find which containers hold more and which hold less. They also report how many objects a container actually holds.

The tricky part about using nonstandard units when measuring capacity is that the more precise the pourable substance is, the more difficult it is to use nonstandard units to measure the amount. It all relates to the amount of space used in the container. If we use same-size buttons to fill a small cup, we can easily count the buttons—but there is space between the buttons in the container. If we used unpopped popcorn, we will have less space between the popcorn in the container—but there are more pieces of popcorn to count. Likewise, if we use rice, the precision increases because there is less space between the grains of rice in the container—but there is a lot more rice to count. So it goes with sand. Imagine the task of counting each grain of sand that a container holds. In these latter cases, we can move to using handfuls, spoonfuls, and other smaller containers as nonstandard units to measure the capacity.

Say we want to find out the capacity of a specific container using a bottle cap as the nonstandard unit. Using capfuls of water, sand, or rice, the children count the number of times they fill and pour each capful to measure the capacity of the container. Actions such as leveling off the substance in the cap each time add to the precision and accuracy. We can help the children communicate that the container holds X number of capfuls.

Using spoonfuls and cupfuls to measure the capacity of containers serves as a bridge to using standard units. Pouring a substance into a container using standard units such as milliliters, teaspoons, tablespoons, cups, pints, liters, quarts, or gallons allows the children to experience the measurement. Children are more likely to understand that the bucket holds five cups of water when they actually pour five cups of water into the bucket. They see that the bucket is full and recall the number of cups of water it took to fill the bucket. This bridging experience into using standard units of measure is in many ways less complicated than using a graduated cylinder or other container that has tick marks to indicate units. That does not mean we don't want children to use containers with standard unit tick marks and graduated cylinders; it just means that children need opportunities to understand what these measurements actually mean. Therefore, pouring five deciliters of milk into a container that has tick marks for deciliters and shows *5 dL* is productive. The children begin to understand what the tick marks on containers mean as they verify the measurements.

In addition to liquid measurement units there are also units for dry measurement, such as bushels and pecks. Interestingly, one U.S. bushel equals 4 pecks (32 dry quarts, 64 dry pints, or 256 gill). If you convert the pourable substances (liquid or dry measurement) into cubic units, you actually move from working with capacity to working with volume. For example, one U.S. gallon equals 231 cubic inches.

To experience volume in a hands-on way, young children can construct three-dimensional shapes using cubes. If children use interlocking cubes to construct a rectangular prism, they can count the number of cubes to determine the space the rectangular prism takes up—which gives the volume of the rectangular prism. *Volume* refers to the amount of substance itself, while *capacity* pertains to a given container.

Just like measuring length and weight, deciding which tool is best for measuring capacity also involves knowing which unit is optimal for each situation. What if we need to measure the capacity of a thimble, paper cup, mug, bucket, or swimming pool? Inviting children to think about which units are small enough or large enough to measure any given object encourages them to apply problem solving and reasoning to the measurement situation.

Area and Perimeter

Area and perimeter are interrelated, yet distinct, branches of measurement. *Area* refers to the number of square units in a region. *Perimeter* is the distance around the outside of a two-dimensional shape or three-dimensional figure. Perimeter is connected to length. Area is connected to length and volume. To find the area, we need to use the length and width to determine the amount needed to cover up the space, whereas when we find volume we need the third dimension (height) along with length and width to determine the measurement.

In many primary classrooms, there is a place where the whole class gathers. Often this place is denoted by a rug or carpet. Let's use the carpeted meeting place to discuss area and perimeter. Perimeter means distance around. If you start at one place on the outside edge of the carpet and walk completely around the carpet on the edge, you have walked the perimeter of the carpet. If you want to find the area of the carpet, you need to use square units to cover the carpet (as in Figure 2–10). The perimeter of the classroom carpet is fourteen units and the area of the carpet is twelve square units. We find these measurements by counting the squares to find the area and counting the sides of the squares that are part of the outside edge to find the perimeter.

The relationship between area and perimeter is further revealed as we think about another carpet that has the same area, but is arranged differently (see Figure 2–11). The area of this classroom carpet is twelve square units; however, the perimeter is no longer fourteen units, but sixteen units. Because of how the square units are arranged, the perimeter of the carpet is different. The perimeter of the classroom carpet can be even greater, if the square units are arranged in the manner shown in Figure 2–12. The area of the carpet is twelve square units, but the perimeter of this carpet is twenty-six units.

Encouraging young children to think about and experiment with these ideas is beneficial. Even young children can position square tiles or cards to find the area of an object, and they can count the outside edges to find the

Figure 2–10
*Finding Area by
Covering the Carpet
with Square Units*

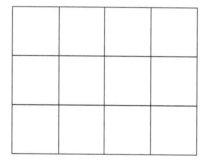

Figure 2–11
*An Example of a
Carpet That Has the
Same Area but a
Different Perimeter*

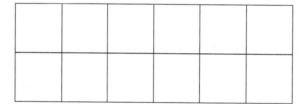

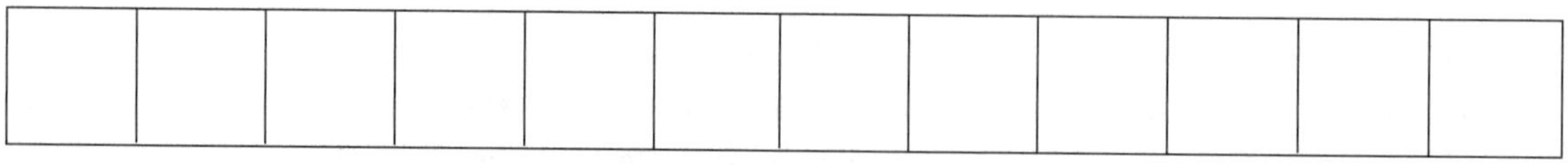

Figure 2–12 *A Carpet with the Same Area but a Still Greater Perimeter*

perimeter. As teachers we can help students become more familiar with these concepts by using the ideas in everyday routines. When we want students to sit on the edge of the carpet, we should invite them to sit on the perimeter. We can construct grid lines on our carpet that indicate square units. If we want the students to sit in rows on the carpet, we can ask them to sit in individual square units. It is important for young children to hear and experience these measurement concepts to build foundations in mathematics.

As with other branches of measurement, the beginning stages of area and perimeter involve comparison. Working with two shapes, children can find which has the greater area, which has the greater perimeter, or if the shapes have equal areas or perimeters. They can directly compare the surface of two objects by putting one on top of the other or *superimposing*. They can cover one shape with smaller shapes and then use the same smaller shapes to try to cover the comparison shape.

As children compare the areas and perimeters of more than two shapes or figures, they can begin to order by size. For example, pictures are ordered from smallest to largest perimeter. Ordering shapes and objects by perimeter or area is a complex task. Sometimes students compare by overlapping or superimposing the shapes and objects. Sometimes they use the same number of smaller shapes on each comparison shape, deciding how much space is left uncovered. Typically simple shapes are easier to compare and order, while intricate shapes (that may include fractional parts of units and concave sides) are more difficult to compare and order by area and perimeter.

Nonstandard and standard units are sometimes used during the comparison and ordering stages of measuring area and perimeter because to find out which shape has the greater area or perimeter, many children place tiles on the shape and then count the number of tiles they used. When they do so, they are using units to measure. The more experiences they have with tiling and counting, the more they begin to see that they may not need to count every square unit to find the area. They may use skip counting, repeated addition, or other strategies. Standard square units involve standard length measurements—square centimeters, square inches, and so on.

Sometimes children use other shapes when working with area and perimeter. For example, instead of using squares, children may use pattern block hexagons as in Figure 2–13.

Using the hexagon as the area unit, the design has an area of seven units. Using the length of one side of the hexagon as the perimeter unit, the design has a perimeter of eighteen units. While it is mathematically possible to report the area and perimeter of this design in square units, we can also discuss the area using the hexagon as long as we define the units. As children work with

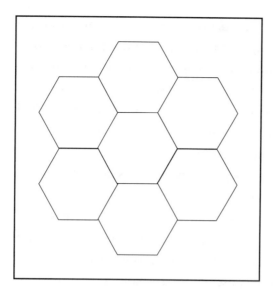

Figure 2–13 *Area and Perimeter with Hexagons*

area and perimeter, they come to understand that certain shapes allow for the entire area to be covered without gaps or spaces between the shapes. The goal is to give the children many opportunities to learn that area involves covering up the space and perimeter involves the distance around the space.

In addition to tiles, blocks, cubes, and grid paper, the geoboard also serves as a great tool for measuring area and perimeter. Because the pins are arranged in a way that makes it easy to show squares within shapes, children can find the whole square units and even partial square units within shapes formed on the geoboard. Deciding which tool is best for measuring area or perimeter involves understanding which unit is optimal for any given situation. What if we need to measure the area or perimeter of an index card, notebook paper, poster board, bulletin board, or billboard? Inviting children to think about which units are small enough or large enough to measure the area or perimeter of any given object encourages them to apply problem solving and reasoning to the task.

Angles

An angle is a geometric figure that has two line segments (or *rays*) that have a common endpoint called the *vertex*. The standard unit of measure for angles is degrees. The number of degrees indicates how open the angle is. Essentially, measuring angles involves thinking about a circle (see Figure 2–14).

While prekindergarten through second graders do not typically learn how to measure angles in degrees using protractors, there are some ways to begin to

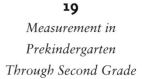

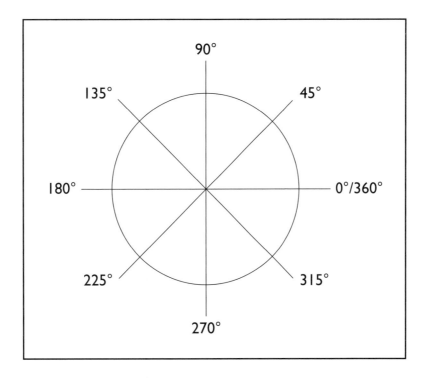

Figure 2–14 *Thinking About Angles by Visualizing a Circle*

introduce the idea of measuring angles with young children. Additionally, some children are ready to learn about angles and we need to have the knowledge to provide them with extended and accelerated learning opportunities.

When comparing two unequal angles, children notice which angle is wide and which angle is narrow, as shown in Figure 2–15. As one first grader shared with me: "Angles are like your mouth. If you are hungry, you open wide. If you are not hungry, you only open your mouth a little bit." This is a great way to think about angles. It helps the children focus on the size of the angle rather than the size of the line segments. When young children compare and order the size of angles, the task is less complicated when the line segments are equal in length, the differences in angle size are more obvious, and the angles are positioned in the same direction.

Protractors are not the only tools that can be used to measure angles. Any square object can help children find out if the angle is greater than, less than, or equal to a right angle (90°). Even the hands on a clock face can serve as the line segments of an angle. Children can identify the size of angles in terms of comparison to each other or in comparison to right angles.

Angles are part of everyday life and we need not avoid instruction on angles just because the children are young. However, the instruction needs to be appropriate and accurate.

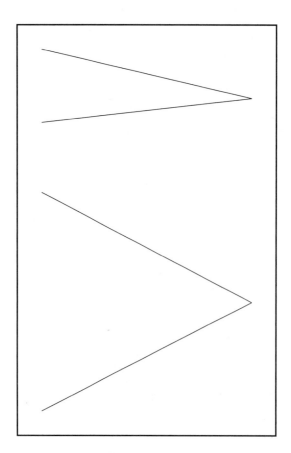

Figure 2–15 *Comparing Two Angles*

Temperature

Temperature is the extent of warmth or coldness of something. Scientifically it is also the measurement of the average kinetic energy, typically reported in degrees. We can determine the temperature of liquids, solids, air, and even ourselves! Children experience temperature every day. They may have a glass of cold milk, wash their hands in warm water, or feel hot when they play outside on a summer day.

While we understand temperature through our experiences, sometimes people exaggerate the temperature. If the room is hot, they may exclaim "I am boiling in here!" If the room is cool, they may say "I am freezing in here!" These inflated temperature statements do not help children understand temperature, because in most situations the statements are erroneous.

Children need to experience temperature and use accurate words to describe the temperature. The beginning stages of measuring temperature involve comparison. Children may compare today's outdoor temperature with yesterday's outdoor temperature. They may compare the temperature of two different cups of juice. They can touch the juice and decide which one is warmer or cooler or if the temperatures feel the same. They may feel rocks found in the sunlight versus rocks found in the shade. Using words like *hotter*, *hottest*, *warmer*, *warmest*, *cooler*, and *coolest*, they can compare the temperature of objects through their sense of touch.

Other senses help us determine temperature as well. We may use our eyes to watch the effects of heating, such as when the sun shines on the snow and it begins to melt. We can use our sense of hearing as we listen to the sounds of food cooking. We can taste ice cream or warm soup. Some smells may trigger thoughts of temperature, such as the smell of a campfire. Each of our senses plays a part in helping us understand temperature, yet it is our sense of touch that gives us immediate information about the temperature of something.

After comparing the temperature of two objects, children are ready to order more than two objects by temperature. Given several cups of water, children can decide how warm or cold each cup of water is in relation to the others. The differences in temperature are more profound if we add various amounts of ice or warm water to the cups. The children touch the water and place the cups in order by increasing or decreasing temperature.

Although there are no specific nonstandard units to use in the measurement of temperature, children sometimes use thermometers to determine temperature without utilizing the scale of degrees. They notice which way and how quickly the red fluid travels. They notice how much of the red fluid shows when the temperature is hot versus cold. All of these observations build foundations for understanding why units for measuring temperature are needed and how they are used.

The way a typical bulb thermometer works involves expansion. The thermometer contains liquid, which is often dyed red. This liquid changes its volume according to the temperature because liquids take up less space when they are cold and more space when they are hot. Because the tube is narrow, we can easily see the expansion of the liquid. Years ago, the fluid used in thermometers was mercury, but because of the toxicity of mercury and the difficulties disposing of it when the thermometer breaks, most classrooms have thermometers that contain dyed alcohol. Reading a bulb thermometer is like reading a vertical number line. Simple thermometers have scales that are easier to read, such as counting by ones, with each tick mark present. More complex thermometers include scales that show skip counting.

In addition to bulb thermometers, there are other tools for measuring temperature. Electronic (digital) thermometers use a computer or other circuit to measure temperature. Spring thermometers and bimetallic strip thermometers use the expansion of metal to measure temperature. The common standard units of measure used by each of these types of thermometers are degrees Celsius or degrees Fahrenheit. Celsius is based on the centigrade scale, where the freezing point of water is 0°C and the boiling point of water is 100°C. In the Fahrenheit scale, the freezing point of water is 32°F and the boiling point of water is 212°F. Therefore the boiling point of water and the freezing point of water are 100°C or 180°F apart. There are other scales used for temperature such as the Kelvin scale and the Rankine scale. Often scientists use these latter two scales because they give a greater range. These scales have no negative numbers and are based upon the point at which all molecular motion stops—*absolute zero*, which happens to be –459.67°F.

Several years ago, at a family math night event, I observed a first grader working with his father using thermometers to check the temperatures of various cups of water. Each time the child reported the temperature, he announced the letter "F" or the letter "C." I was quite impressed with his ability to read and report the temperatures and asked him if he knew what the *F* and the *C* meant on the thermometer. "Yes!" he proudly declared, "It means *Fahren-hot* and *Fahren-cold*!"

Time

Time is one of the most significant, yet abstract, concepts for young children. There are durations of time, specific times, and events in time. Time can feel long or short depending on what you are doing. In addition to all the different aspects of time, children often receive mixed messages about time. When a student requests something of the teacher, the busy teacher may respond "I'll be there in a second." But the teacher doesn't actually mean one second. She really means *I will be there when I can*, which will most likely be much longer than one second. Or when something is in disarray, the teacher may announce to the class, "Wait a minute!" The teacher most likely wants the class to stop and think, which may or may not take one minute. Perhaps the child is with a parent waiting in a long line at the grocery store. The annoyed parent may declare, "This is taking hours!" when really it has only been ten minutes. While not intended to confuse children, our misuses and exaggerations of time actually add to the difficulties some children have with time.

Comparing the time of day specific events occur is a great way to help young children compare time, particularly when the events are significant to the children. If we ask children to think about when they eat lunch compared to when they go home from school, they can use words like *before* and *after* or *earlier* and *later* to describe the time of the events. Daily discussion about the class schedule helps the students compare time in a meaningful context. They also begin to understand ideas such as morning, afternoon, evening, and night and they can describe events (e.g., breakfast, recess, dinner, and bedtime) that occur during those times.

Children can also compare days of the week using *yesterday*, *today*, and *tomorrow*. If today is Tuesday, we can compare it to yesterday (e.g., Monday comes before Tuesday and Tuesday comes after Monday) or tomorrow will be Wednesday because it is the day after Tuesday. We can also compare months and years the same way. Given two months in the same year, April and May, we can talk about which month comes before or which month is later in the year. Many teachers address these comparisons with their classes during *calendar time*. During this activity, children use the calendar to discuss the day and to compare days, weeks, months, and years.

Comparing durations of time is also helpful. Say we compare how long it takes us to build a simple tower with blocks compared to how long it takes us to eat lunch. We do not need to use units of time to compare these two events because they can occur simultaneously. When the tower is built and we've only just begun to eat our lunch, we understand that these two events are unequal durations of time. It takes longer to eat lunch than it takes to build a simple tower out of blocks. Does it take longer for the class to line up than it does for the teacher to say the alphabet? Can you wash your hands for the same amount of time that it takes to sing a nursery rhyme about a twinkling star? Providing opportunities for young children to compare durations of time is beneficial.

Ordering time involves the same ideas as comparing time, there are just more events, seasons, days, weeks, months, or other durations of time to compare and place in order. Given five daily events (lunch, math, reading, packing up to go home, and recess), the children can order these events according to which events happened before and after the others. Children can order the days of the week and the months of the year. When ordering these it is important to remember that there is a continuous cycle. Weeks follow weeks and the twelve-month cycle is followed by other twelve-month cycles. Night follows day but also comes before the next day.

Durations of time can also be placed in order. Given three tasks (write your name, jump up and down fifty times, and clean your desk), the children can work together, engaging in the tasks simultaneously to find out which tasks take longer or shorter amounts of time. The tasks can then be placed in order of the time it takes to complete them—the greatest amount of time to the least amount of time or vice versa.

There are several types of nonstandard units that can be used to measure time. If we want to know how long it takes you to tie your shoe, we can use nonstandard units to measure this specific duration of time. While you tie your shoe, we can clap our hands, recite a poem, say the alphabet, walk around the classroom, perform jumping jacks, or any other task. It make take you twenty-five claps to tie your shoe—or two readings of the poem, three and one half times through the alphabet, one walk around the classroom, or forty-two jumping jacks. These are just some of the many nonstandard units that can be used to measure durations of time.

Children also use daily occurrences to measure durations of time in nonstandard units. A first grader once told me, "It is going to take me about four TV shows to read this book." To make this statement, he used his background experiences (the lengths of TV shows) to estimate how long it was going to take him to finish reading the book.

After watching an hour-long performance at a schoolwide assembly, I asked a class of kindergartners how long they thought the performance was. The responses included long, too long, and kind of long. But one child shared, "It lasted about two recesses." This student used a familiar duration of time (thirty minutes) as a nonstandard unit to measure a new duration of time (the assembly)!

Telling time using clocks is an exciting step for young children. They are often fascinated with clocks and make attempts to read time even at early ages. Some children wear digital watches and appear to be able to tell time, but many are just reading the digits rather than understanding what time it actually is or how that time relates to the rest of the day. Using analog clocks helps children understand time, especially when they see the clock as circular number lines (one number line for hours, one for minutes, and one for seconds).

Learning to tell time involves understanding the lengths of standard units of time. Children need to know the lengths of minutes, seconds, hours, and fractions of these standard units. Ask a group of young students to cover their eyes and raise their hands when they think one minute has passed; it is amazing to see the wide range. Some children raise their hands right away, others wait only a few seconds. In many classrooms, all the children have raised their hands by about thirty seconds—unless the students have developed counting strategies to estimate time. Asking students to experiment with activities that take about one minute to complete is a great way for them to recognize the length of one minute. We can do the same for one second and fractional parts of an hour.

The sequence of teaching how to tell time using clocks begins with time by the hour. We want children to understand what the clock looks like on the hour. We also want them to understand which hour it is. While they are learning about time on the hour, it is also beneficial to talk about before the hour and after the hour. Which o'clock will it be next? Which o'clock just past? Time by the hour is the foundation of telling time using clocks.

Time by the half hour is typically next in the sequence. At this point children are looking for half past something or something-thirty. One of the things that poses problems for some children when reading or showing half past the hour with an analog clock is the position of the hour hand. When setting or drawing a clock, some children want the hour hand to point directly to the hour. But this is not how a properly functioning analog clock looks. On the half hour, the hour hand should be exactly *between* the hour that just passed and the hour that is next. Children who have difficulties with this concept often have not had enough experiences with on the hour, before the hour, and after the hour.

Telling and understanding time by the quarter hour is next in the sequence. For many children this concept is simplified by thinking in terms of fractional parts of a circle and fractional parts of sixty. Quarter past two or 2:15 is the same as ¼ past the hour of 2:00. Half past two or 2:30 is the same as ½ or ²⁄₄ past the hour of 2:00. Likewise, quarter of three or 2:45 is the same as ¾ past the hour of 2:00 or ¼ before the hour of 3:00. Notice how before and after the hour is still critically important to telling and understanding time by quarter hours.

Time by five-minute intervals and then time by one-minute intervals are next in the sequence of learning to tell and understanding time using an analog clock. Children use skip counting and chunking as they learn to read and understand the time using analog clocks. Again the concept of before the hour and after the hour plays a major role in accuracy. When it is 2:55, the hour hand is not as close to the 2 as it is to the 3 because it is almost, but not yet, the 3:00 hour. It is 55 minutes past the hour of 2:00. Helping children frame each time between the relative two hours is beneficial.

Typically, after children learn how to tell and understand time by the hour, half hour, quarter hour, five-minute intervals, and one-minute intervals, we shift gears and focus on elapsed time. If you have ever taught elapsed time, you know that it is a concept that many children struggle with. Older children try setting up equations to subtract and add time—but they are often confused trying to work in base-ten to make sixty minutes equal one hour. Some of the students are perplexed about what elapsed time actually means. Elapsed time can be a frustrating thing for some students and their teachers.

One of the ways I have found success with teaching elapsed time is to teach it throughout the sequence. When children learn to tell time by the hour, we can teach elapsed time by the hour. When children learn to tell time by the half hour, we can teach elapsed time by the half hour. When they learn time by the quarter hour, they can learn elapsed time by the quarter hour—and so on. The idea of teaching elapsed time throughout the process of learning time intervals is very powerful because it not only helps the children understand elapsed time more completely, but it also puts time in meaningful context. Our busy lives are full of implementing elapsed time—teaching children to problem solve and apply the standard units of measuring time through elapsed time helps children solidify and own the knowledge.

Using digital clocks while children learn how to tell and show time on analog clocks is also important. We do not need to teach digital time separately. By intersecting the teaching of analog and digital we actually give the students more ways to communicate time. The most efficient way to record time is to write it with digits, which is exactly how the digital clock looks. We should use both digital and analog clocks when we teach students how to tell and understand time.

When teaching time, I prefer to use analog clocks that have geared hands. These clocks allow children to show accurate times. When students use clocks that are not geared, they tend to show erroneous times—such as the hour hand pointing exactly to the 5 and the minute hand pointing exactly to the 6 to show 5:30. There are also geared clocks that have a space underneath the clock that is a dry-erase board. Students can write the time with digits that match the time shown on the clock face. Clocks of all different sizes should be used. Other tools such as stopwatches, sundials, water clocks, and atomic clocks should also be used in addition to mechanical clocks.

Deciding which tool is best for measuring time also involves knowing which unit is optimal for the situation. What if we need to measure the time duration of a handshake, a meal preparation, a marathon, a plant's growth, or a sea voyage around the globe? Inviting children to think about which units are small enough or large enough to measure any given duration of time encourages them to apply problem solving and reasoning to the measurement situation.

Whether we are working with length, weight, capacity and volume, area and perimeter, angle, temperature, or time, we are measuring. But measurement is not limited to these branches alone. We can also measure electrical current, earthquakes, light years, sound, and even computer hard drive space. The more advances we make in science and technology, the more measurable attributes we will continue to find—the math continues to grow!

Applying Appropriate Techniques and Formulas in Measurement

All in good measure

There are a host of techniques and formulas that can be used to determine measurements. As children explore with comparing, ordering, and using nonstandard and standard units, they come to understand that certain techniques are more accurate and efficient than other techniques. When using nonstandard or standard units, the measurement process is essentially the same for any branch of measurement. You need to decide which unit to use and then use some kind of technique that involves comparing the unit to the object being measured. *Iterating* is a common technique—"The number of units can be determined by *iterating* the unit (repeatedly laying the unit against the object) and counting the iterations or by using a measurement tool" (National Council of Teachers of Mathematics 2000, 105).

Another technique students use to measure is showing the units with their fingers or hands and then iterating this measurement. I refer to this technique as *unitizing*. They start with a specific size of unit and then make motions as they visually move this unit along the length or other measurable characteristic of the object. For some people this technique is helpful; for others it is not so accurate because it is difficult to consistently keep the unit aligned during the process.

When students begin to use operations in their measurement techniques, they are applying techniques that serve as the basis for formulas. A formula is essentially an equation that shows a mathematical relationship. When a student measures one side of the square at two inches—then skip counts "2, 4, 6, 8" or adds $2 + 2 + 2 + 2$ or multiplies 4×2—the student is actually using the formula for the perimeter of a square $p = 4s$ where p = perimeter and s = the measurement of one side. If the students covers a rectangle with 12 one-inch tiles and adds the rows $(4 + 4 + 4)$ or multiples the number of rows by the number of tiles in each row (3×4), the child is using the formula for the area of a rectangle $a = lw$ where a = area and l = length and w = width.

Some young children are even ready to use the formula for the area of a triangle $(a = \frac{1}{2} bh)$ because they understand that a triangle's area is half that of a square with the same width (base) and length (height). There are even some young children who ask questions that are related to sophisticated formulas. For example, while working with geoboards and trying to figure out the lengths of the sides of a right triangle, some children begin to ask questions like, "What about the long side? How can we use what we know to figure it out? How are the lengths of the sides related?" These questions about the hypotenuse and the relationship between the lengths of the sides of a right triangle are related to the Pythagorean Theorem ($a^2 + b^2 = c^2$).

Using Estimation in Measurement

Using estimation is a critical aspect of measurement. Essentially, all measurements are actually approximations at various levels of precision. With this in mind, measurement is not possible without estimation.

Estimating is more than guessing. Good estimations involve some understanding of the measurement context and the units of choice. This is one of the reasons why we encourage students to use referents. A *referent* is a benchmark that serves as a unit for estimating. If a group of students is trying to figure out how many cookies are in a jar, we can show them a smaller jar with the same type of cookies in it. We tell them how many cookies are in the smaller jar so they can use that information to estimate how many cookies are in the larger jar. Referents do not have to be smaller. The referent could be a larger jar. The point is that the referent gives information that can be used in estimating. The more students estimate, the better they become at using referents—and they actually learn to create their own referents within larger quantities.

Another important aspect of estimation is using *range-based techniques*. "Estimation should not be used as a strategy to find the correct answer. Rather estimation should involve using mathematical skill to predict information within an adequate range" (Taylor-Cox 2001, 213). *Range-based techniques* help students learn how to estimate within a range, rather than focus on a lucky guess that happens to be the exact answer. We want to teach children to think about what the range should be and where their estimate falls within the range. In this way, estimation helps us know if our measurements make sense.

Measurement is an exciting part of mathematics. Finding and reporting the size, amount, or degree of any object requires the real-life application of estimation, number sense, counting, operations, equations, and geometry. We need to provide many opportunities for our students to help them understand the measurable attributes of objects and how to use appropriate techniques and formulas to determine those measurements.

CHAPTER 3

Targeted Instruction in Measurement

Index Question for Weight

Which is heaviest? How do you know?

Ms. Vanderbilt teaches a class of twenty-one prekindergarteners. All of her students have diverse needs and various levels of understanding mathematics. To gain information about what her students know about measuring weight, Ms. Vanderbilt uses a simple informal assessment. While the children work at math centers, Ms. Vanderbilt calls children to join her at the back table. She displays a simple balance and two plastic eggs. Unbeknownst to the students Ms. Vanderbilt placed two small batteries in one egg (a blue egg) and nothing in the other egg (an orange egg). Ms. Vanderbilt places one egg on each side of the balance and asks each child which egg is heavier. Ms. Vanderbilt helps the children record their responses.

See:
Number, p. 53
Algebra, p. 19
Data, p. 32
Geometry, p. 32

Explanation of the Weight Index Question

The children have explored with the balances for about a week, but Ms. Vanderbilt has not officially started teaching the measurement of weight to her students. She simply placed several balances in the explore center, allowing the children to begin to learn about the properties and characteristics of this measurement tool.

Ms. Vanderbilt wants to find out if and to what degree the students know how to compare the weights of objects. She is looking to see if the students know how to use the balance and if they understand what the information means. She shows them a real balance with real objects on it in Figure 3–1. She also photographed the balance with the eggs on it and printed this photograph to allow the children the opportunity to use the photograph in their responses. Ms. Vanderbilt knows that the photograph is not necessary because she could have the children simply look at the real balance and then respond. But Ms. Vanderbilt is trying to help her students learn how to record information and she was concerned that they would focus too much on sketching a balance and not enough on the math—so she printed out copies of the photograph for students to use in their responses.

Identifying which egg is heavier is a way for the children to measure by comparing weight. The follow-up question, "How do you know?" encourages the children to explain and prove their math thinking.

Student Responses

Calissa's Response

See:
Number, p. 53
Algebra, p. 19
Data, p. 34
Geometry, p. 34

Calissa successfully identifies the blue egg as the heavier egg, as shown in Figure 3–2. To further verify her answer, she draws an arrow that points to the heavier egg on the scale. Calissa has a solid understanding of how to use the balance to compare weight. However, Calissa's answer to the how-do-you-know question is minimal—*Bigger.* The blue egg is not bigger because the eggs are the same size. Yet Calissa's answer indicates that she knows something about comparing measurements because she uses the comparative ending *er.* Several other children in Ms. Vanderbilt's class gave similar responses. They were able to identify the heavier egg, but they were not able to explain their answer. Ms. Vanderbilt grouped these children with Calissa for on-grade-level instruction. It is exciting to know that these children know how to use a simple balance. They are ready for some instruction on using reasoning and proof to provide explanations.

Aylin's Response

Aylin accurately responds that the blue egg is heavier (see Figure 3–3, p. 33). She also draws a line and a small circle, which seems to serve as Aylin's way of identifying the heavier egg. She has a similar small circle drawn beside her answer. It appears that Aylin is attempting to code her answer in some way. In response to the how-do-you-know question, Aylin writes *Down.* This is really an excellent answer. Aylin's proof that the blue egg is heavier is that the balance is unequal because the heavier egg is down and the lighter egg is up.

Name ____________________

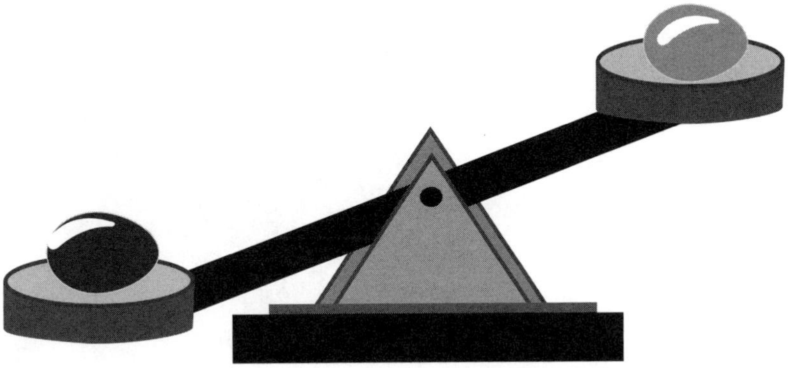

Which egg is heavier?

How do you know?

Figure 3–1 *Weight Index Question*

Name _Calissa_

Which egg is heavier?

blue

How do you know?

Bigger

Figure 3–2 *Calissa's Response*

The answer is really impressive considering Aylin is in prekindergarten and the class has not yet focused on measuring weight. Clearly Aylin and the other students who gave similar responses are ready for a challenge. They do not need to participate in a lesson on identifying which is the heavier or lighter object on a simple balance. They need to work with more complex measurement ideas.

Amyla's Response

Amyla's response in Figure 3–4 indicates that she does not yet know how to use the balance. It appears that Amyla wrote some random letters as a response to

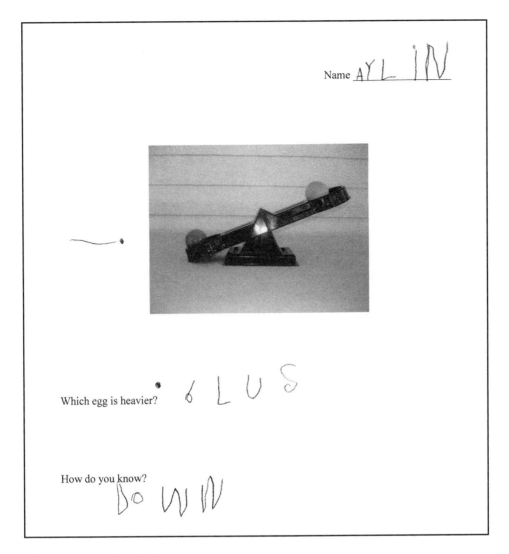

Name A Y L I N

Which egg is heavier?   b L U S

How do you know?   D o W N

Figure 3–3 *Alyin's Response*

the index question. But Ms. Vanderbilt knows that Amyla has been working in the writing center learning how to write her friend's name (Prada). These letters *P*, *R*, and *A* are some of the letters in her friend's name. So the letters are important to Amyla even though they are unrelated to the measurement index question.

Amyla and the other students who were not yet able to accurately identify the heavier egg will work with Ms. Vanderbilt building foundations with measuring weight.

Using a simple index question allowed Ms. Vanderbilt to form groups based on student needs. The preassessment was brief and Ms. Vanderbilt was able to do a quick sort of the indexes and form her purposeful groups for tomorrow's instruction. Now Ms. Vanderbilt can focus her attention on

Name A M Y L A

Which egg is heavier?

P RAP

How do you know?

Figure 3–4 *Amyla's Response*

adjusting the levels of cognitive demand and targeting instruction aimed at meeting the students' academic needs.

Small Group, Targeted Instruction

See:
Number, p. 56
Algebra, p. 25
Data, p. 37
Geometry, p. 37

There were three distinct levels revealed in the responses given on the pre-assessment Ms. Vanderbilt gave to her prekindergarteners. Ms. Vanderbilt plans to meet with each group to target instruction while the rest of the students in the class work in small groups at centers or work on independent activities.

The First Group Works with Weight

The middle group (including Calissa) appears to have a solid understanding of weight and how to use a simple balance to compare weight. All of the children in this group were able to identify the heavier egg. Ms. Vanderbilt has this group focus on using math thinking to support answers. They do not need to spend time learning how to use a simple balance to identify the heavier of two objects because they already have that knowledge. Ms. Vanderbilt compacts this information and concentrates instruction on comparing and ordering the weights of more than two objects as a way to encourage the students to use math reasoning and measurement vocabulary.

Ms. Vanderbilt organizes the students into dyads and triads. She gives each small group a balance and four plastic eggs that have pennies in them. Each set of plastic eggs has one green egg with twelve pennies in it, one blue egg with six pennies in it, one pink egg with six pennies in it, and one purple egg with three pennies in it. Ms. Vanderbilt asks the students to find out which egg weighs the most without opening the eggs. The students begin using the balances to compare the weights of the plastic eggs. After a few minutes, all of the students find that the green egg weighs more than the other eggs. Ms. Vanderbilt asks the students to explain how they found out which egg weighs more. One of the students says, "The balance goes down." The other students agree. Ms. Vanderbilt asks, "Why does the balance go down?" One student replies, "The green egg is more." Ms. Vanderbilt asks, "How do you know it is more weight?" Another student responds, "It is bigger." Ms. Vanderbilt holds up a green egg and a blue egg, asking "Is the green egg longer?" The students reply, "No." Ms. Vanderbilt asks, "Is the green egg wider?" The students respond, "No." Ms. Vanderbilt explains, "If it is not longer or wider or deeper, then it can't be bigger." One student says, "But it is more heavy." "Yes," replies Ms. Vanderbilt, "it is *heavier*." The children repeat the word, "Heavier." Ms. Vanderbilt asks, "The green egg is heavier than what?" The children reply, "Heavier than the blue egg." Ms. Vanderbilt asks, "What can we say about the blue egg?" One student responds, "The blue egg is not heavier." "Right," says Ms. Vanderbilt, "the blue egg is not heavier than the green egg. The blue egg is *lighter* than the green egg." The students repeat the word, "Lighter." Ms. Vanderbilt asks, "Which egg will make the balance go down?" The students reply, "Heavier." Ms. Vanderbilt asks, "What happens to the lighter egg on the balance?" A student responds, "It goes up," as he shows the motion with his balance. The other students show the action on their balances as well.

Ms. Vanderbilt continues, "I want you to try using the words *heavier* and *lighter* as you compare the blue, green, pink, and purple eggs." The students begin working on the task. Ms. Vanderbilt encourages the students to use the terms *heavier* and *lighter* as they compare the weights of the eggs. Ms. Vanderbilt also helps the students to complete the measurement comparisons by asking, "Heavier than what?" and "Lighter than what?"

"Now I want you to put the eggs in order from heaviest to lightest," announces Ms. Vanderbilt. The students begin ordering the eggs by weight. One of the students says, "We have a problem." Ms. Vanderbilt responds, "What is the problem?" The student answers, "Some of the eggs are not heavy or light." Another student says, "Yeah. We have some that are the same, too." Ms. Vanderbilt asks, "How can we describe it if the eggs weigh the same?" Another student responds, "Equal." The other students chime in, "Equal!" Ms. Vanderbilt encourages the students to use math words and ideas as they work on the task.

After the students order the eggs by weight and find the two eggs that weigh an equal amount, Ms. Vanderbilt asks the students to write and draw about the activity in their math journals. In this lesson, the students learned how to use words and ideas to explain and prove their math thinking as related to measuring weight because Ms. Vanderbilt targeted the instruction to meet their academic needs.

The Second Group Works with Weight

The responses given by this group of students (including Aylin) indicate that they already understand how to compare weight and give simple explanations of their thinking. In essence, this group's knowledge of comparing weight is advanced. Ms. Vanderbilt meets with this group to provide targeted instruction through forward-mapping to measuring weight with more complex tools.

Ms. Vanderbilt displays several measurement tools. She has a simple balance, a pan balance, and a bucket balance. She asks the students to explore with the balances using various objects (bottle caps, crayons, staplers, and large rocks) to find out how the balances are alike and different. After some exploring the students dictate how the balances are similar and different to Ms. Vanderbilt, who records the information on chart paper. The students share that all of the balances have two places to put objects. They also share that while all of the balances go down when the objects are heavier, some of the balances only go down a little bit. Some of the balances can fit more objects and some of the balances can only fit a few objects.

Because the students are able to compare and contrast the balances so easily, Ms. Vanderbilt introduces another measurement tool. She displays a platform scale. Ms Vanderbilt asks the students to observe what happens as she places a stapler on the platform scale. The students are surprised to see the arrow on the dial move. Ms. Vanderbilt asks, "What do you think this means?" One student replies, "I think that is how heavy the stapler is." Another student adds, "Yes. Me too." Another student says, "The stapler weighs almost one." Ms. Vanderbilt explains, "The dial shows us how many pounds objects weigh." A student responds, "The stapler weighs almost one pound!" The other students agree. "I wonder how many pounds this rock weighs," says a student as he picks up the heavy rock. "Let's weigh it!" exclaim the students. Ms. Vanderbilt helps the students place the rock on the scale. Everyone watches the arrow on the dial rapidly move to show the weight. The students announce, "It weighs four pounds!"

Ms. Vanderbilt asks the students to work with their partners to weigh objects on the platform scales and record the information by drawing pictures, recording numbers, or writing words in their math journals.

The Third Group Works with Weight

These students (including Amyla) gave responses that indicate that they are not yet able to use a simple balance to compare weight. Ms. Vanderbilt uses scaffolding and activating prior knowledge with this group. She will help the students connect the new learning experiences to their earlier experiences with measuring length. Ms. Vanderbilt uses the Three Bear Family™ counters as the objects to be weighed because of the connections between height and weight. In the set, each small bear is 1 inch tall and weighs 4 grams, each medium-sized bear is 1¼ inches tall and weighs 8 grams, and each large bear is 1½ inches tall and weighs 12 grams. The focus of the lesson will not be these standard units of measurement because the students are not ready for that. Instead the focus will be on which bears are heavier and which bears are lighter. Comparing the weight of the bears is simplified because the weight of the bears is scaled according to height.

Ms. Vanderbilt organizes the students into dyads and triads and gives each group two bears (one large and one small) that are the same color (red). She asks the students to place one bear on one side of the balance and the other bear on the other side of the balance. The children do so and observe the results. Ms. Vanderbilt asks, "What happened?" The students describe the position of the balances. Ms. Vanderbilt asks, "Why do you think that happened?" One of the students replies, "The big one is more." The other students agree. Ms. Vanderbilt says, "Yes. The larger bear is *heavier*. Everyone point to the *heavy* bear on your balance." The children point to the heavier bear. Ms. Vanderbilt asks, "Which bear is not heavy?" The students point to the lighter bear. Ms. Vanderbilt explains that this bear is *lighter* and asks, "What happens to the lighter bear on the balance?" The students respond, "It goes up." Ms. Vanderbilt affirms the students' answer and provides them with an additional (medium-sized) bear, saying "Try comparing the weight of this bear to your other bears." The students begin comparing the weights of the three bears.

After the students compare the weights, Ms. Vanderbilt gives each group another set of three bears—but these bears are not red, they are yellow. Ms. Vanderbilt asks the students to compare the weights of the red bears to the yellow bears. She helps the students describe which bears are *heavier* and which bears are *lighter*. She also helps the students use the word *equal* to describe bears that weigh the same. Ms. Vanderbilt gives the students additional bears and invites them to explore further and record information in their math journals.

In this differentiated mathematics lesson, Ms. Vanderbilt targets the instruction for each group. She uses tiering because each of the groups works with weight, but the activities are tiered to the appropriate levels. The tasks are differentiated by level, in that one group focuses on ordering by weight, one group focuses on using more complex measurement tools, and the other group works with comparing weight. Ms. Vanderbilt also uses back-mapping, forward-mapping, scaffolding, compacting, and activating prior knowledge with the groups.

Index Question for Capacity

About how many cubes will fit in the cylinder? How did you make your estimate?

See:
Number, p. 61
Algebra, p. 26
Data, p. 40
Geometry, p. 41

Ms. Hollins teaches kindergarten. She has twenty-two students in her class. Ms. Hollins' students have studied how to measure and estimate length and weight and are now beginning a unit on capacity. Ms. Hollins uses a simple preassessment to gather information about what her students know about measuring capacity and estimation. She gathers the students in a circle and displays a cylinder with one cube in it (see Figure 3–5). Ms. Hollins asks, "How many cubes will fit in the cylinder?" The children think about their responses. Ms. Hollins adds, "After you answer, I want you to think about how you made your estimate." She explains to the students that during the day, each person will have a chance to go to the back table to look at the cylinder and answer the questions. The students know that during center time or independent work time, they can look to see if the table is open so they can have their individual turns to respond to the index question.

Figure 3–5

Explanation of the Capacity Index Question

The index question Ms. Hollins uses (see Figure 3–6) involves both an assessment of what the students have learned and an assessment of what the students will learn. It serves as an ongoing assessment and a preassessment. The students have learned about estimation, so Ms. Hollins will check to see if they can apply what they know about estimation to this new idea of capacity. The how-many question is framed by the referent that is given (one cube is in the cylinder). The "How did you make your estimate?" question reveals more information about how the children used estimation to answer the question.

Name _____

Use words, pictures, and numbers to answer the questions.

About how many cubes will fit in the cylinder?

How did you make your estimate?

Figure 3–6 *Capacity Index Question*

Student Responses

Mecklit's Response

Mecklit's response in Figure 3–7 indicates that she knows a good deal about estimating capacity. Ten is a great estimate for the number of cubes that will fit in the cylinder. The actual amount of cubes that will fit in the cylinder is twelve, making ten an estimate that is within a good range of the actual answer. Mecklit's response to the "How did you make your estimate?" question is "Because I was thinking." Although it is great that Mecklit was thinking and that she can write that she was thinking, the explanation does not give any support to her answer. Mecklit is not yet able to explain how she estimated the capacity or give reasoning and proof to support her estimate.

Mecklit and the other students who gave similar responses are ready for some on-grade-level instruction. They have solid knowledge about estimating

See:
Number, p. 61
Algebra, p. 26
Data, p. 41
Geometry, p. 41

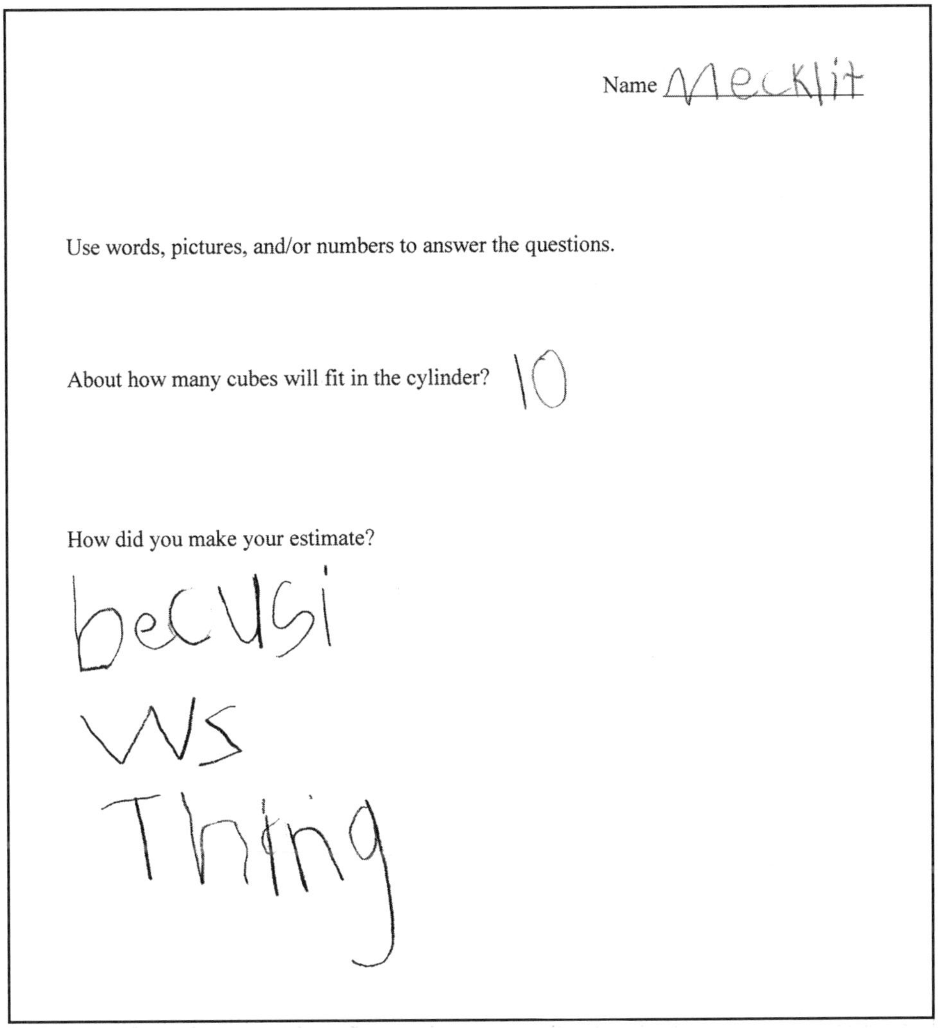

Name Mecklit

Use words, pictures, and/or numbers to answer the questions.

About how many cubes will fit in the cylinder? 10

How did you make your estimate?

becusi ws thing

Figure 3–7
Mecklit's Response

capacity, but need to focus on supporting their estimates with mathematical thinking. Ms. Hollins formed a group with these students to target instruction.

Karolina's Response

The response Karolina gives is excellent (see Figure 3–8). Karolina estimates that ten cubes will fit in the cylinder. The estimate is very good because it is within a good range of the actual number of cubes that will fit in the cylinder. Karolina's illustration of a cylinder with ten cubes in it serves as a way to explain and support her estimate. She also writes *Becusse it look cindur big and cindur small*. Karolina read her writing in this way: "Because it looks like the cylinder is big and the cubes are small." Her words offer a good description of how ten cubes can fit in the cylinder. Ms. Hollins groups Karolina with the other students who gave responses that indicate that they understand how to make good estimates related to capacity. They do not need on-grade-level instruction. They need to be challenged at the next level.

Daven's Response

Daven estimates that forty-one cubes will fit in the cylinder in Figure 3–9. His estimate is too high to be considered within a good range. To explain how he made his estimate Daven writes *I okati*, which means "I'm OK at it." Daven believes he is OK at estimating—and with some further instruction, he will be! Daven and the other students who gave similar responses will work with Ms. Hollins on building foundations in estimating capacity.

Ms. Hollins' use of the index question allowed her to reveal the different levels of understanding of her students and to form groups based on student needs. The assessment took each student only a few minutes to complete and did not take away from Ms. Hollins' instructional time because the students worked on this task independently. It took Ms. Hollins a couple of minutes to do a quick sort of the indexes and form her purposeful groups for the next day's instruction. Now Ms. Hollins can focus her attention on adjusting the levels of cognitive demand and targeting instruction aimed at meeting the students' academic needs.

Small Group, Targeted Instruction

See:
Number, p. 64
Algebra, p. 31
Data, p. 44
Geometry, p. 46

There were three separate levels revealed in the responses given on the pre-assessment Ms. Hollins gave her kindergarteners. Ms. Hollins plans to work with each group to differentiate the instruction. While Ms. Hollins teaches each group, the other groups explore with measurement manipulatives or write

Name Karolina

Use words, pictures, and/or numbers to answer the questions.

About how many cubes will fit in the cylinder? 10

How did you make your estimate?

becusse it look cinduv big and cinduv smoll

Figure 3–8 *Karolina's Response*

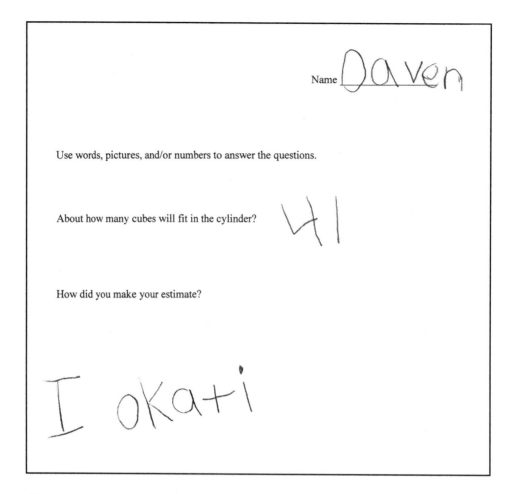

Figure 3-9 *Daven's Response*

in their math journals. Ms. Hollins will let the students know when it is time to switch groups (approximately every ten minutes).

The First Group Works with Capacity

The middle group (including Mecklit) appears to have a solid understanding of how to estimate capacity—although these students need further experiences with estimating capacity for them to explain and give reasons to support their estimates. Ms. Hollins will use scaffolding and activating prior knowledge to teach the on-grade-level capacity objectives. The students will learn more about estimating using nonstandard units of capacity. She plans to have the students estimate the capacity of two different containers. The containers Ms. Hollins uses are somewhat complex. One of the containers is short, but very wide; the other container is tall, but very narrow. Unbeknownst to the students, both of the containers have the same capacity.

Ms. Hollins displays the two containers and a bucket of sand and says, "How can we find out how much these containers hold?" One student says, "It is not the same as saying how many cubes will fit." Ms. Hollins asks, "How is it different?" The student replies, "Because you can't count sand!" The other students agree. Ms. Hollins suggests they use scoopfuls to measure the sand. The students agree with the idea. Ms. Hollins asks, "How many scoopfuls of sand do you think will fit in each of the containers?" The students estimate various amounts. Ms. Hollins gives each student a piece of paper and asks the students to write down how many scoopfuls they think will fit in the tall container and how many will fit in the short container. The students share their estimates. Using a funnel, Ms. Hollins pours the first scoopful into the tall container and asks, "Does anyone want to change their estimate?" No one in the group wants to change their estimate. Ms. Hollins invites one of the students to pour another scoopful of sand in the tall container. With Ms. Hollins' help, the students take turns pouring in scoopfuls of sand. After each scoopful Ms. Hollins asks the group what the total is so far and if anyone wants to change their estimate. It is not until the container is close to halfway full (four scoopfuls) that some of the students decide to change their estimates. Ms. Hollins encourages them to record their adjustments and asks them to explain why they decided to adjust their estimates. One of the students says, "Well, I thought it was going to be four and now I know four is not enough." Another student says, "Me too. I am changing to eight." Ms. Hollins responds, "Tell us why you are adjusting your estimate to eight." The student replies, "Because I can see it is in the middle." Another student replies, "I get it. You think it is four and four more." Some of the students count on their fingers to verify this statement and decide to change their estimates also.

Ms. Hollins continues to help students pour in scoopfuls, keep count, and discuss the reasoning that supports adjustments to estimates that are made. The students learn that the container holds nine scoopfuls of sand. Because they were invited to adjust their estimates and explain their reasoning throughout the process, all of the students were successful.

Ms. Hollins organizes the group into smaller groups and gives each group a bag of sand, a box top to use as the work area, a scoop, and a short container (just like the one she originally displayed to the group). She explains to the students that they will estimate, pour scoopfuls, and adjust estimates using this short container (just like they just did with the tall container). Ms. Hollins reminds the students that when they adjust the estimate, they need to share the reason why they are making the adjustment. The students begin working with their partners on the task while Ms. Hollins takes a quick walk around the room to check on the other students.

When Ms. Hollins rejoins the group, she provides support and guidance as the students work on estimating and measuring the capacity of the container. They are excited to learn that the short container also holds nine scoopfuls of sand, just like the tall container. Ms. Hollins invites the students to go back to their desks and write about this activity in their math journals.

The Second Group Works with Capacity

These students (including Karolina) gave responses that indicate that they already know a lot about estimating capacity and explaining their reasoning. Because they understand at a level beyond the rest of the class, they are ready for a challenge. Ms. Hollins will compact the instruction related to nonstandard units and forward-map to teaching standard units for measuring and estimating capacity.

Ms. Hollins gives each of the students a measuring cup and asks them to investigate it and share what they notice. One student shares, "There are numbers on it." Another student adds, "It looks like you could fill it up." "Or part way up," adds another student. Ms. Hollins responds, "Yes. This is a measuring cup. It holds three cups. See if you can find the line that shows one cup." The students look for the number *1* on the cup and point to the location. Ms. Hollins asks, "What else do you notice about the measuring cup?" One of the students replies, "It can hold two cups if you fill it to here [pointing to the location]." "Or you can fill it up for three cups," adds another student.

Ms. Hollins organizes the students into dyads and gives each group a bag of sand and a box top. She asks them to use their measuring cups to show two cups of sand. The students work together showing the accurate amount. Ms. Hollins asks them to show one cup, and then three cups of sand; they do so successfully.

Ms. Hollins shows the students an empty bucket and asks them to estimate how many cups of sand it will take to fill the bucket. The students discuss their estimates which range from ten to twenty. Ms. Hollins helps the students pour a full measuring cup into the bucket and asks, "How many cups of sand do we have in the bucket so far?" The students reply, "One." Ms. Hollins says, "Are you sure? How many cups fit in the whole measuring cup?" The students realize that they will need to count by threes as they fill the bucket with sand.

When the bucket is almost full at fifteen cups of sand, Ms. Hollins asks, "Will any more sand fit into the bucket?" The students decide that the bucket will not hold three more cups, but it may hold one more cup. The students add one more cup of sand to the bucket. One of the students says, "I think it could hold just a little bit more. Is there anything smaller than a cup?" Ms. Hollins responds, "Do you have to use a whole cup?" "No," says one of the other students, "we could use half of a cup." The other students agree and the group places half a cup of sand into the bucket. The bucket is full. The students announce that the bucket holds 16½ cups of sand.

Ms. Hollins shows the students another bucket that is somewhat bigger than the bucket they just filled. She asks the group to estimate how many cups of sand will fit into this new bucket. The students record their estimates and explanations in their math journals.

The Third Group Works with Capacity

This group of students (including Daven) gave responses that indicate that they need additional support with estimating capacity. Ms. Hollins targets the instruction for this group of students by back-mapping to comparing capacities (measurement without using nonstandard units). She starts with the same two containers that she used with the first group.

Ms. Hollins displays the containers and the bucket of sand and asks, "Do you think one of these containers holds more sand than the other container?" The students attempt to compare the containers by predicting capacity, but there is some disagreement about how much the containers will hold. Ms. Hollins picks up the tallest container and says, "Let's try this one first." Several students agree with the selection because they believe it will hold the most. Ms. Hollins asks, "How will we figure out how much the container holds?" One student says, "We have to pour sand in it." Ms. Hollins invites him to use a funnel and scoop to pour sand in the container. The rest of the students observe. When the container is full, the students cheer. Ms. Hollins asks, "How can we describe the container?" One student says, "It is full." The other students agree. Ms. Hollins holds up the short container and says, "How about this container?" One student replies, "It is not full." Ms. Hollins replies, "Yes. It is not full. We can also say it is *empty*."

Ms. Hollins says, "Turn to the person beside you and discuss what you think will happen if we pour the sand from the full container into the empty container." The students begin discussing what they think will happen. Some of the students think the sand will spill out. Other students think there is not enough sand. Ms. Hollins asks each student to share what they think will happen and why they think that will happen. Ms. Hollins helps the students use capacity words and ideas in their explanations.

Ms. Hollins helps the students pour the sand from the tall container to the short container. The students are surprised to see the result. One student says, "Hey, it all fits." Another student says, "The short one is full." Another student adds, "Now the tall one is empty. Let's see it again!" Ms. Hollins helps the students pour the sand from one container back into the other container several times.

Ms. Hollins shows the students a third (larger) container and asks the students to estimate if this new container will hold more or less than the other two containers. The students estimate that the new container will hold more than the other two containers. Ms. Hollins wants the students to record their estimates so she asks them to draw the new container and how much sand would be in it, if they were to pour in all the sand from one of the other containers.

In this differentiated mathematics lesson, Ms. Hollins successfully challenges and supports each of the groups through targeted instruction. Ms. Hollins had many options for how to differentiate the instruction. She used tiering since each of the groups work with estimating capacity using sand, but the activities are tiered to the appropriate levels. The tasks are differentiated by level in that one group focuses on estimating capacity using nonstandard units (scoopfuls), one group focuses on using standard units (cups and half cups) to estimate and measure capacity, and the other group works with estimating capacity by comparing. Ms. Hollins also uses back-mapping, forward-mapping, scaffolding, compacting, and activating prior knowledge with the groups.

Index Question for Time

What time is it?

See:
Number, p. 69
Algebra, p. 32
Data, p. 49
Geometry, p. 52

Mr. Levin has a class of twenty-five first graders. The students range in ability and mathematics expertise. Mr. Levin's class has been learning how to tell and understand time using analog and digital clocks. Mr. Levin wants to know where his students currently are in their levels of understanding time. He developed an assessment aimed at revealing the different levels of knowledge of the students.

Explanation of the Time Index Question

Mr. Levin wants to give an assessment to the students to find out what they currently know about time on an analog clock (see Figure 3–10). While this assessment may appear simple, it is actually somewhat complex because there are several right answers and several wrong answers. The wrong answers are similar to the errors children commonly make when learning how to read time on an analog clock. Mr. Levin thinks that there may be some students who

What time is it?

Name _____

Erika says, "It is 6:00."

Trevon says, "It is half past 12."

Maria says, "It is 1:30."

Sukey says, "It is a few minutes after 6."

Ashley says, "It is 12:30."

Xavier says, "It will be 1:00 in 30 minutes."

Davey says, "It is 1:30."

Which students are right?

Figure 3–10 *Time Index Question*

See:
Number, p. 69
Algebra, p. 34
Data, p. 51
Geometry, p. 52

can find all of the correct answers. For these students, he will differentiate the assessment by asking them what time it will be in fifteen minutes (to see if they know elapsed time and telling time by fifteen-minute intervals).

Student Responses

Cindy's Response

Cindy's response in Figure 3–11 indicates that she knows about time by the half hour. To indicate which students know the correct time on the clock on the index question, Cindy writes *yes* or *no* next to each answer. This is a clever way to help her keep track of her answers. Cindy knows that the time on the clock is 12:30. She also knows that it will be 1:00 in thirty minutes. However, Cindy does not yet know that this time can also be described as *half past 12:00*. Cindy has a solid understanding of time by the half hour, she just needs some targeted instruction in the area of *half past*. Mr. Levin learned that other students found some, but not all, of the accurate ways to tell a specific

Figure 3–11
Cindy's Response

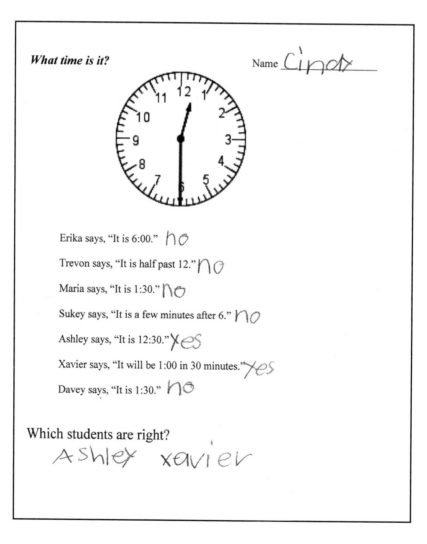

time by the half hour. He grouped these students with Cindy to provide differentiated instruction.

Canaan's Response

Canaan's response in Figure 3–12 is completely accurate. He crosses off the
incorrect answers and places a ring around the correct answers. Noticing that
Canaan and a couple of other students completed the index question accurately and quickly, Mr. Levin took the opportunity to provide an extension to
the preassessment for these students. He asked Canaan what time it would be

Figure 3–12
Canaan's Response

in fifteen minutes. Canaan is used to having challenges, so he wrote the answer to Mr. Levin's question on his paper. He also drew a clock to show the time. Clearly Canaan is working above grade level in this area of measurement. Mr. Levin groups Canaan with other students who gave completely accurate responses to the index question and the extended question.

Ray's Response

Ray answers the index question in Figure 3–13 by indicating that the time on the clock is 6:00. While Ray does have a system for keeping track of his answers, he does not yet know how to tell time by the half hour. Mr. Levin will work with Ray and the other students who gave similar responses.

Mr. Levin's use of the index question allowed him to reveal the different levels of understanding of his students. The index question served as an ongoing assessment during a unit on time. It only took the students a few minutes to respond to the index question and just a few more minutes for Mr.

Figure 3–13

Ray's Response

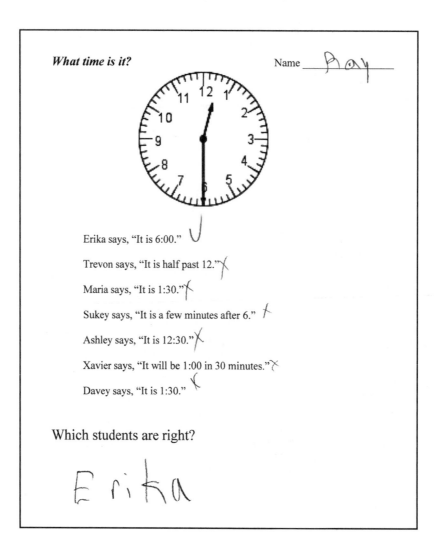

Levin to quickly sort the information to help him form his groups for targeted instruction.

Targeted Instruction in Measurement

Small Group, Targeted Instruction

There were three levels revealed in the responses given on the preassessment Mr. Levin gave to his first graders. Mr. Levin uses tiering as he provides differentiated instruction through targeted mini-lessons in telling and understanding time. While Mr. Levin works with each group, the rest of the class works on completing an assignment from yesterday and making clocks out of paper.

See:
Number, p. 74
Algebra, p. 37
Data, p. 54
Geometry, p. 56

The First Group Works with Time

This group of students (including Cindy) appears to have some understanding of time by the half hour, but the students do not yet know all of the ways to name time by the half hour. With this group, Mr. Levin will use activating prior knowledge, scaffolding, some compacting, and a little bit of backmapping. He will teach the students more about time by the half hour by helping them make connections to fractions and telling time by the hour.

Mr. Levin calls the students to work with him. He initiates a discussion about the length of one hour. The students share specific trips, TV shows, and CDs that are one hour long. Mr. Levin organizes the students into dyads and gives each group a small geared clock and asks them to show 1:00. All of the groups successfully show the time (as Mr. Levin expected). He asks the students to think about what time it will be one hour later. The students are not sure what to do. Mr. Levin explains, while demonstrating the clock movement on his large geared clock, "One hour later means one whole time around the clock with the minute hand." The students mimic his motions on their clocks. "It is 2:00," says one of the students. Everyone agrees. Mr. Levin asks, "What if we start at 8:30. Show me what the clock looks like one hour later." The students show 8:30 and then one whole time around the clock with the minute hand. Mr. Levin says, "No matter what time you start on, one hour later will be one time around the clock." After a few more practices with one hour later, Mr. Levin tries one hour before with the students. They are successful, so he asks them to show 30 minutes later. With some guidance and discussion about ½ and halfway around the clock, all of the students are able to show 30 minutes later. The time shown on their clocks is 11:30. Mr. Levin shares with the students that this time can also be called *half past 11* or *half past 11:00*.

At this time Mr. Levin shows the students a large paper cube that is marked + *one hour*, + *30 minutes*, + *½ hour*, – *one hour*, – *30 minutes*, – *½ hour*. Mr. Levin explains that he will roll the cube and the students will show the new time on their clocks. "Since the time now is almost 10:00," says Mr. Levin looking up at the classroom clock, "let's start with 10:00 on your clocks." Mr. Levin rolls the paper cube and it lands on + *30 minutes*. Mr. Levin says, "Think about whether this time is earlier or later than 10:00. The students work with their partners to show 10:00 plus 30 minutes. Mr. Levin asks one of the groups to explain how they figured out what time to show on their clock. The explanation involves moving the minute hand halfway around the clock. Another student shares, "Yep. We did the same thing." Other students agree. Mr. Levin rolls the paper cube again and it lands on – *one hour*. The students work together to figure out how to show 10:30 minus one hour. After showing the time one of the students says, "It is one hour earlier." Another student adds, "You just go backward all the way around one time." The other students agree.

Mr. Levin rolls the paper cube a few more times, helping the students show and explain the times. Then he gives each dyad a smaller version of the paper cube and encourages the students to continue the activity. One person rolls the cube, the other person shows the time, they both check the time, and then they switch roles. During this time, Mr. Levin can work directly with the couple of students who need extra support with this activity.

The activity that Mr. Levin introduces to the students can now become a center activity because the students understand the directions and procedures well enough to work on this by themselves. Later when Mr. Levin includes this as a center activity, he adds one other component. He wants the students to draw and write the times as they take turns with the activity because he wants to monitor the progress and to know if there are any misconceptions he needs to address. Another great thing about this activity is that it can now be easily differentiated simply by changing the labels on the paper cubes.

The Second Group Works with Time

Mr. Levin focuses his attention on the next group. The students in this group (including Canaan) need a challenge because they already know how to tell time by the hour, by the half hour, and by fifteen-minute intervals. The students already know what Mr. Levin's curriculum guide indicates that he should teach. Therefore Mr. Levin needs to use forward-mapping to teach these students how to tell time by five-minute intervals and how to work with elapsed time.

Mr. Levin gives each of the students a geared clock and asks them to show 4:45. All of the students can accurately show the time. Mr. Levin asks,

"What time will it be in five minutes?" The students set their clocks for 4:50 and announce the time. Mr. Levin asks the students to explain how they figured out the time. One student explains, "I just count on five more minutes." Another student explains, "I used counting by fives." Another student adds, "I know 45 plus five is 50."

Mr. Levin divides the six students into two teams and explains to the students that they are going to learn a new game. This game is called the *earlier or later game*. Mr. Levin sets two large geared clocks on the chalkboard ledge. He has the two teams stand in front of the clocks. He sets the clocks for 3:00 and explains that we will announce how much time later or earlier and the first team to set the clock accurately gains one point. The members of each team can work together to show the time. When they believe they have set the clock correctly, they need to sit down. When everyone on the team is sitting down and Mr. Levin can see the clock, he knows that the team is finished. The first elapsed time Mr. Levin announces is "Twenty minutes later." The groups gather around their respective clocks, discuss the time, and everyone on the team sits down to show that they are finished. The first round is a tie, so Mr. Levin says each team gets ½ of a point. The clocks stay on 3:20 and Mr. Levin announces, "Thirty minutes before." The groups work to show the time. It takes a little bit longer than the first round, but one of the teams accurately shows the time before the other team. The game continues for eight more rounds. Mr. Levin varies the difficulty of the elapsed time for each new round. He even includes earlier and later by hours and minutes. The content is extremely high, but the students are motivated and have their teammates with whom to figure out the answers. As it turns out, the game ended with a score of four to four. The students appreciated the game so much that they decided to play again at recess. Needless to say, Mr. Levin was thrilled when the students asked to borrow the clocks to use during recess.

The Third Group Works with Time

Mr. Levin works with the third group (including Ray). He will use back-mapping to help the students build foundations in understanding time by the hour. Mr. Levin invites the students to work with partners to show times on geared clocks. He announces, "Show 8:00," and encourages the students to talk about the time as they show it on the geared clocks. The students highlight the placements of the hour and the minute hand for the specific time. After Mr. Levin announces a few more on-the-hour times and the students discuss and show these times, Mr. Levin asks, "What do you notice about all the times we have shown so far?" One student says, "All the times are o'clock times." Another student says, "All the times have the long hand on the twelve."

"And the short hand points to whatever hour it is," says another student. The other students agree. Mr. Levin says, "You are right. We have shown several o'clock times or on-the-hour times. Now we are going to try something else. I will show you a time on the clock and your job is to tell me which o'clock time has just passed and which o'clock time will be next." The students are ready for the task.

Mr. Levin shows 2:30. The students talk with their partners about the previous hour and the hour that will be next using the placements of the hour hand and the minute hand. One of the students raises her hand to respond. When Mr. Levin calls on her she says, "It used to be two o'clock and three o'clock is coming up next." Mr. Levin says, "Exactly! How do you know?" The student responds, "Because the hour hand is halfway between the two and the three." The group agrees and tries a few more half-hour times. The students discuss each time and Mr. Levin calls on different students to answer. They work hard to explain the hour that just passed and the hour that is next. Mr. Levin makes this activity into a game. Each dyad is shown a time and asked to name the previous hour and the following hour. The dyads receive a point for stating the correct answer. An extra point is given if the students can tell the time. Mr. Levin intentionally shows all half-hour times. He knows that the students will gain more accuracy in reading half-hour times by focusing on the previous hour and the next hour. One of the students shared, "I know it is 9:30 because it used to be 9:00 and next it will be 10:00." Another student says, "It is halfway past 9, that's why the little hand is halfway between 9 and 10." Mr. Levin is elated with the progress the group has made. The students are also pleased. They ask Mr. Levin if they can play the game again tomorrow.

In this differentiated mathematics lesson, Mr. Levin successfully challenges and supports each of the groups through targeted instruction. Mr. Levin had many options for how to differentiate the instruction and chose to tier the lesson, use back-mapping, use forward-mapping, scaffold, compact, and activate prior knowledge. Each of the groups worked with different levels of telling time. Additionally, the students worked with appropriate levels of elapsed time.

Index Question for Length and Other Measurable Attributes

See:
Number, p. 77
Algebra, p. 40
Data, p. 57
Geometry, p. 59

What are the ways you could measure the rectangle?

Ms. Brown teaches a class of twenty-three second graders. Her students have various levels of understanding of measurement. The students have worked with measuring lengths using rulers. They have not yet worked with other aspects of measurement. Ms. Brown wants to see where her students are in their understanding and accuracy of measuring length. She also wants to see if any of the students know other measurable attributes.

Explanation of the Length and Other Measurable Attributes Index Question

Ms. Brown wants to know her students' current levels of understanding of length and other measurable attributes (see Figure 3–14). Because the students have worked with rulers, this is an assessment of their level of accuracy and their understanding of units and how to record measurements. The assessment also includes the opportunity for students to show if they know any other measurable attributes that apply in this situation.

Student Responses

Maya's Response

Maya gives a solid response in Figure 3–15 to the index question. She explains, "You can measure the rectangle in inches. You can measure the rectangle in centimeters." She is right. You can measure the rectangle in inches and in centimeters. Her ability to use a ruler accurately is also apparent because the length of the rectangle is $4\frac{1}{2}$ inches and approximately $11\frac{1}{2}$ centimeters. Maya's response is correct. However, Maya does not include any other measurements. She does not measure the width of the rectangle and she does not include any other ways to measure the rectangle. Maya is working on grade level. Ms. Brown forms a small group with Maya and the other students who know how to use the ruler to accurately measure length in inches and centimeters.

Karen's Response

Karen's response (see Figure 3–16) includes a lot of information. She writes that you can measure the rectangle "using a ruler" and "you can measure the rectangle using cups." Karen's idea of measuring the rectangle with cups is not connected to capacity, but to using cups as nonstandard units for length. She attempts to illustrate equal spacing in the lengths of her cups and records that the rectangle is 8 cups in length. She also measures the length of the rectangle with the ruler and accurately records that the rectangle is $4\frac{1}{2}$ inches long. When she measures the length of the rectangle in centimeters, she inaccurately records that the rectangle is 11 centimeters (she is off by approximately $\frac{1}{2}$ centimeter). Karen also measures the width of the rectangle. She accurately

See:
Number, p. 78
Algebra, p. 41
Data, p. 58
Geometry, p. 60

Name _____

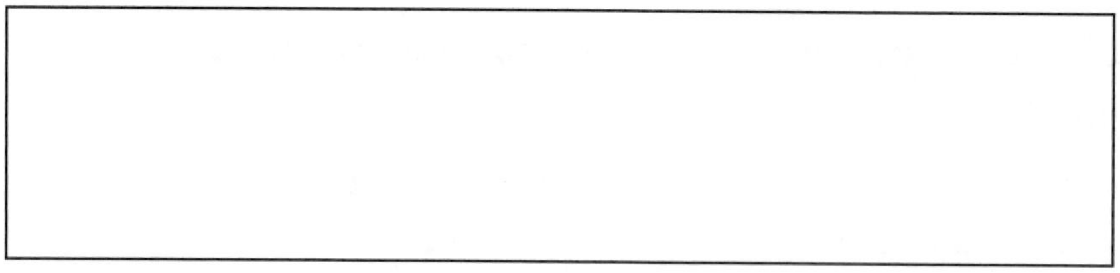

What are the ways you could measure this rectangle?

Measure the rectangle and record your findings.

Figure 3–14 *Length and Other Measurable Attributes Index Question*

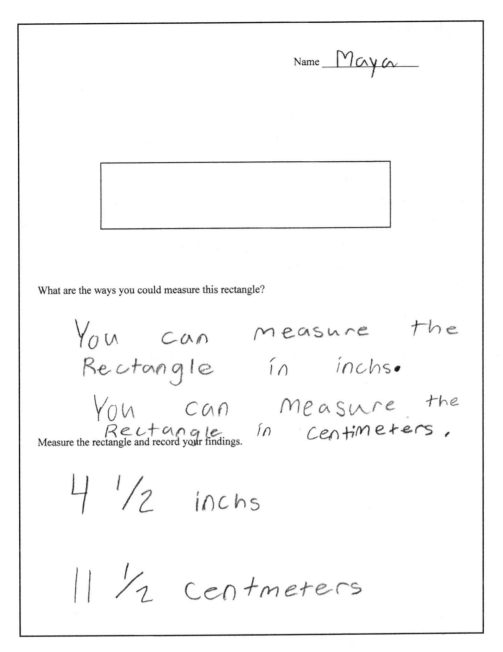

Name Maya

What are the ways you could measure this rectangle?

You can measure the Rectangle in inchs.

You can measure the Rectangle in centimeters.

Measure the rectangle and record your findings.

4 1/2 inchs

11 1/2 centmeters

Figure 3–15 *Maya's Response (student artwork reduced)*

records that the width of the rectangle is 1 inch, but incorrectly records that the width of the rectangle is 3 centimeters (she is off by approximately ½ a centimeter). Ms. Brown knows that Karen often finds the correct answer, but does not always write the correct answer. Ms. Brown decides to group Karen with other students who gave above-grade-level responses to the index question. If Ms. Brown sees that Karen does not belong in this group, she will simply move her to another group.

Nathaly's Response

Nathaly's response, shown in Figure 3–17, indicates that she needs more instruction in measurement. She does write that the rectangle could be measured with a ruler and records 4½, but does not include the unit she used. She does not measure with centimeters, nor does she attempt to measure any other attributes of the rectangle. Ms. Brown groups Nathaly with other students who gave similar responses.

The index question Ms. Brown used provided her with valuable information. Based on the evidence, Ms. Brown adjusted the level of cognitive demand and targeted instruction for three groups of students. She knew that the students working at a high level needed challenge and the students who did not yet understand much about these concepts needed additional support and scaffolding.

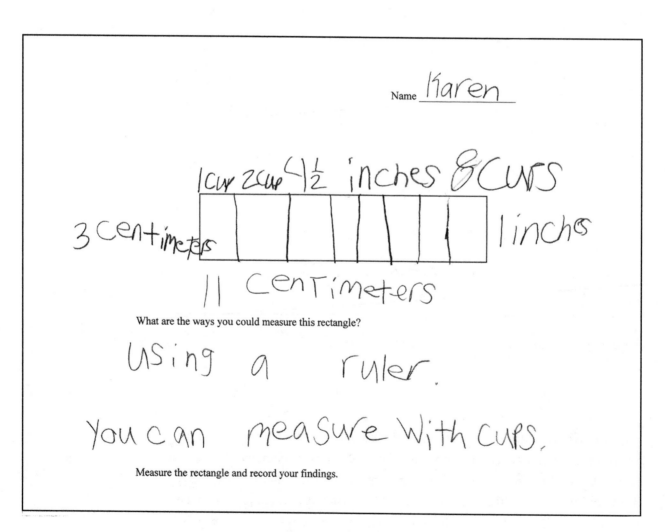

Figure 3–16 *Karen's Response (student artwork reduced)*

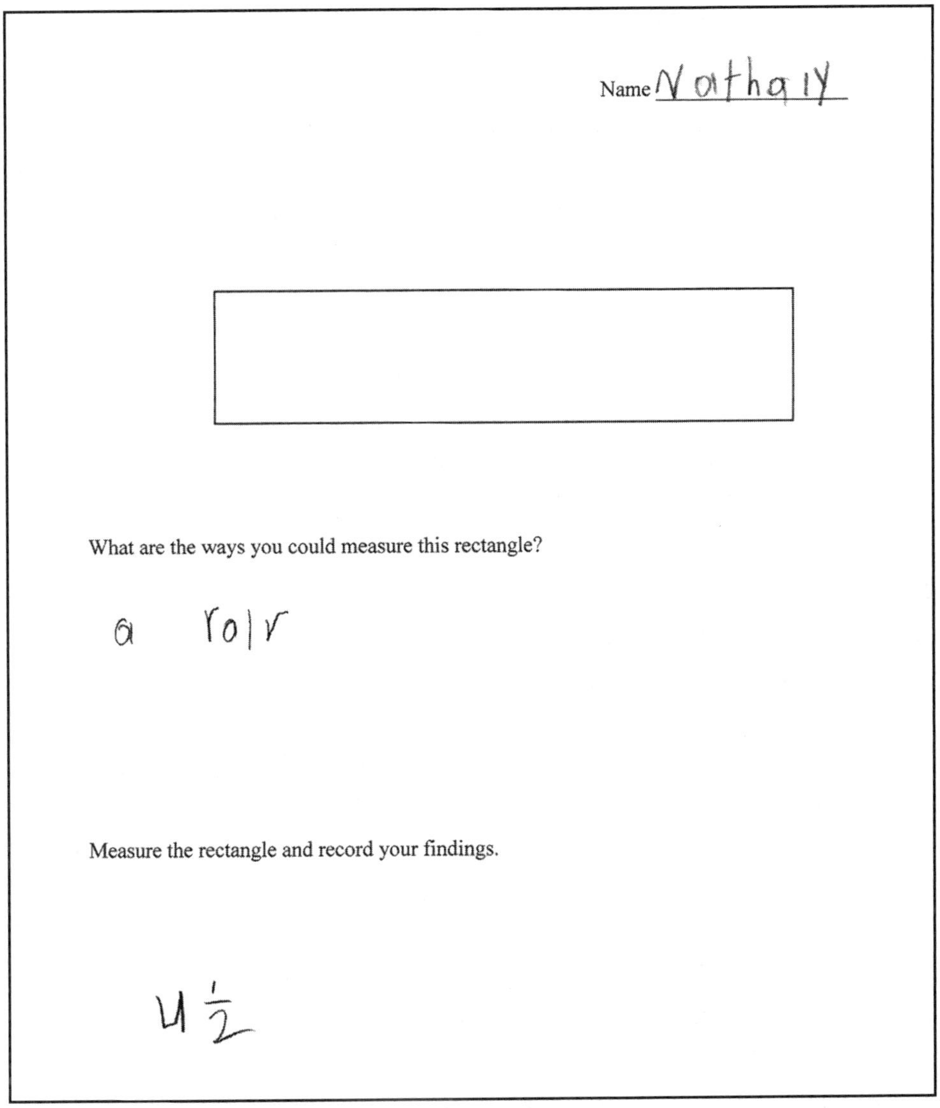

Name Nathaly

What are the ways you could measure this rectangle?

a rolr

Measure the rectangle and record your findings.

$4\frac{1}{2}$

Figure 3–17 *Nathaly's Response (student artwork reduced)*

Small Group, Targeted Instruction

There were three levels revealed in the responses given on the preassessment Ms. Brown gave to her second graders. Ms. Brown uses tiering to provide differentiated levels of instruction for the groups. While Ms. Brown meets with each group to target instruction, the rest of the class works on completing differentiated assignments from the previous day and differentiated center activities.

See:
Number, p. 81
Algebra, p. 46
Data, p. 62
Geometry, p. 62

The First Group Works with Length and Other Measurable Attributes

This group of students (including Maya) works with Ms. Brown on the on-grade-level objectives for measurement. She can compact some of this information because the students know how to use a ruler to measure length in inches and centimeters. But their knowledge is not complete because they do not yet understand that the width can also be measured. Ms. Brown plans to provide opportunities to work with this important measurement idea.

Ms. Brown works with the first group of students. She organizes the students into dyads and distributes various sizes of rectangles. Ms. Brown asks the students, "How can we measure these rectangles?" One student answers, "We need rulers." Ms. Brown invites him to go get some rulers for the group to use. Another student says, "We can put the rectangles in order by size." Ms. Brown responds, "Tell us what you mean." The student compares two rectangles by height, saying "See, this one is bigger than the other one." Ms. Brown responds, "Yes, it is taller. Is there another way you can measure the rectangle?" Another student responds, "You could see how fat it is." Ms. Brown says, "How would you measure that?" The student uses a ruler to measure the width of one of the rectangles. Ms. Brown says, "You are measuring the width of the rectangle." Another student picks up a square rectangle and says, "How do you know which is the width and which is the length?" Ms. Brown responds, "Length is from end to end and is usually the longer part." "But what about this square? All the sides are the same size," says the student. Ms. Brown replies, "With a square, the length and the width are both the same size."

Ms. Brown asks the students to work with their partners to measure the dimensions of the rectangles. They are invited to write the measurements on sticky notes and check each other's measurements. As the students work on this task, Ms. Brown monitors their progress, assists when necessary, and encourages the students to talk about the measurements and strategies as they work.

Because the students completed the task so accurately and efficiently, Ms. Brown knew it was appropriate to do a little forward-mapping with the group. She asked the students to look at one of the rectangles and think about the measurable attributes. The students talked about length and width. But Ms. Brown propelled the students' thinking further by asking, "What else could be measured?" One of the students answers, "Is there a way to measure the corners?" The other students like his idea. Ms. Brown responds, "Yes. We can measure the angles." The students ask, "How do you do that?"

Ms. Brown shows the students a right angle template. One of the students says, "It looks like two rulers put together." Another student adds, "Yeah. One ruler goes up and the other goes sideways." Ms. Brown explains to the students that they can use a right angle template to check and see if an angle is a right angle or not. The students use the new tool to check the rectangles and discover that all of the rectangles have right angles. Ms. Brown explains that the measurement of a right angle is 90°.

Ms. Brown asks, "Are there any shapes that you can think of that do not have right angles?" The students think about this question. One of the students says, "What about triangles?" Ms. Brown says, "Let's use the triangles in the tangram set to check for right angles." One of the students goes to get several sets of tangrams. The students work with their partners using right angle templates to check for right angles. "Can we check other shapes, too?" one of the students asks. "Absolutely!" answers Ms. Brown. While checking for right angles the students announce which shapes have 90° angles and which shapes do not have 90° angles. One of the students asks, "Can we measure the lengths and widths of these shapes?" "Of course," responds Ms. Brown. As the students measure, Ms. Brown helps them find the base and height of each shape.

In this learning experience, the students gained a great deal of knowledge about measurement. Because they quickly learned about including width as a measurement, Ms. Brown was able to forward-map into more advanced measurement ideas (i.e., measuring angles and measuring the base and height of shapes).

The Second Group Works with Length and Other Measurable Attributes

This group of students (including Karen) has a strong understanding of length and other measurable attributes. The curriculum guide indicates that Ms. Brown should teach the students how to use rulers to measure length and width in inches and centimeters. Yet this group of students already knows how to do so. Ms. Brown will compact the instruction of measuring length and width and forward-map to challenge the students by teaching them about perimeter and area.

Ms. Brown introduces perimeter and area to this group of students using the same paper rectangles she used with the previous group. She organizes the students into dyads and triads and gives each group one rectangle. Ms. Brown asks the students to measure the length and width in centimeters and they do so accurately (as Ms. Brown expected). Ms. Brown asks, "What if a little ant walks around the edge of the rectangle? What is the total distance if the ant goes completely around the rectangle?" The students think about the question and begin measuring. Ms. Brown monitors the students' progress and guides when necessary. Each of the small groups is able to find the total distance around the given rectangle. Ms. Brown shares, "The distance around the outside is called the *perimeter*." The students try out the new word by repeating it. One of the students asks, "Can you measure perimeter with inches, too?" Ms. Brown says, "You can use different units to measure the perimeter just like you can use different units to measure the length and width."

Ms. Brown gives each group of students another rectangle and asks them to find out if the perimeter of this rectangle is smaller or larger than the first rectangle that they measured. The groups work on the task. The students are able to accurately measure each side of the rectangle and find the sum of the measurements to report the perimeter. The students compare the perimeters and share the findings.

Ms. Brown distributes some paper and asks the students to work with their partners to draw a rectangle that has a perimeter of twelve inches. The students use their rulers to work on the task. One of the groups draws a twelve inch line for one of the sides, but then realizes the perimeter includes all four sides. A couple of other groups erase line segments as they work on forming a rectangle that has a perimeter of twelve inches. When the students are finished, Ms. Brown asks them to flip their paper over so they cannot see each other's rectangles. Ms. Brown asks, "Do you think everyone's rectangle looks the same?" The students respond, "Yes." Ms. Brown says, "OK, flip your papers over so everyone can see." The students do so and are surprised to find that there are several different rectangles. They trade papers to check each other's work and find that they have three different rectangles that all have a perimeter of twelve inches.

"There is also another way to measure these rectangles," announces Ms. Brown, "We can find the area of each rectangle." One of the students asks, "What's the area?" Ms. Brown answers, "The area is the number of square units inside the rectangle." One of the students asks, "Do you mean like where the ant lives?" Ms. Brown says, "Tell us more about what you are asking." The student says, "Well, if the ant walks all the way around the rectangle, it is like marking off its own place. So the ant can live inside." Ms. Brown says, "That's a great way to think about it. So if you were the ant, which of these rectangles would you chose?" The students look at the three rectangles. One of the students says, "I would want the rectangle with the most room." Ms. Brown says, "Would you want the rectangle that has the largest area?" "Yes," answer the students. One student asks, "How do we measure the area?" Ms.

Brown reaches for the bucket of square tiles and says, "We find out how many square inches cover it up."

The students begin measuring the areas of the three rectangles. With Ms. Brown's help, they find that even though all three rectangles have the same perimeter, they each have a different area (see Figure 3–18). The students decide that if they were ants, they would prefer the rectangle that has an area of nine square inches.

In this differentiated mini-lesson, the students learned about measuring area and perimeter. Using their knowledge of length and width and the hands-on manipulatives (rulers and tiles), the students were able to work with the above-grade-level concepts with ease. Ms. Brown compacted what the students already knew and focused on advanced material for them.

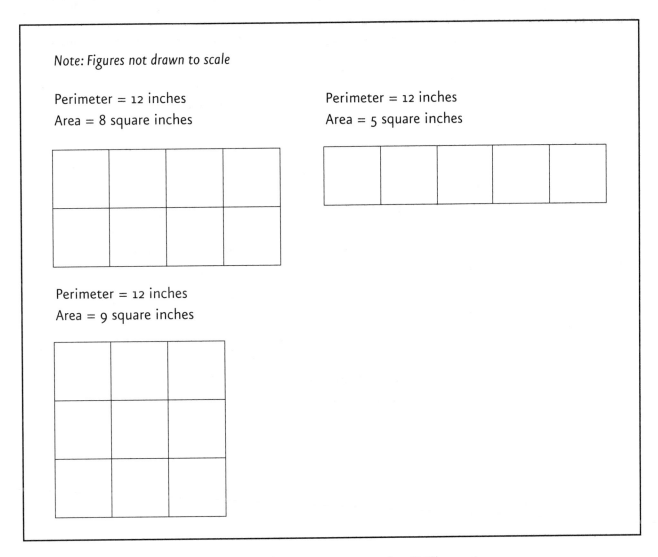

Figure 3–18 *A Set of Rectangles with the Same Perimeter but Different Areas*

The Third Group Works with Length and Other Measurable Attributes

Ms. Brown works with this group of students (including Nathaly) on strengthening their understanding of using and reporting measurement units. Ms. Brown back-maps to using objects to measure standard units as a way to build foundations in using rulers to measure standard units. Ms. Brown also activates prior knowledge and scaffolds by using a variety of measurement tools.

Ms. Brown organizes the students into smaller groups and distributes inch cubes and centimeter cubes. While the students explore with the manipulatives, Ms. Brown asks them questions about the sizes of the cubes. She also asks the students to work together to show given measurements. She announces, "Show seven centimeters," and watches to see which cubes they use and how they line up the cubes. The students soon connect to their prior experiences and are comfortable using the cubes to show given measurements in inches or centimeters.

Then Ms. Brown presents the students with several different types of rulers. Some of the rulers are plastic and some are wooden. Several of the rulers have both inches and centimeters. Some of the rulers only have centimeters or inches. Most of the rulers are twelve inches long. But some are six or eighteen inches long. Ms. Brown guides a discussion about the similarities and differences among and between the rulers. She encourages the students to use the names of the units.

Using the same paper rectangles that she used with the other groups, Ms. Brown asks the students to work with their partners to measure the rectangles using cubes and then verifying the measurements with the rulers. As the students work on the task, Ms. Brown focuses attention on helping them measure and name the units accurately. The students gain greater comfort with the units as they measure the various rectangles using cubes and rulers.

In this differentiated mathematics lesson, Ms. Brown targets the instruction for each group of students. She activates prior knowledge, compacts, scaffolds, tiers, and uses back-mapping and forward-mapping. The mini-lessons are targeted to meet the needs of all of the students.

CHAPTER 4

Grouping Students in Measurement

Using Measurement Vocabulary to Label Small Groups

Using content vocabulary to identify small groups offers a twofold benefit. The group names help with classroom management and offer opportunities for exposure to content vocabulary. Because the groups are always changing, we want the names of the groups to change as well. Using specific content vocabulary as the group names encourages the students to see and hear more math. These are lists of some of the measurement vocabulary that could be used as names of small groups.

See:
Number, p. 98
Algebra, p. 48
Data, p. 67
Geometry, p. 69

Measurable attributes and tools words:

Analog clock	Centigram
Analog scale	Centiliter
Angle	Centimeter
Area	Cents
Balance	Century
Big	Cubic
Bucket balance	Cup
Bushel	Day
Calendar	Decade
Capacity	Decigram

Deciliter

Decimeter

Degrees Celsius

Degrees Fahrenheit

Dekagram

Dekaliter

Dekameter

Digital clock

Digital scale

Dime

Dollars

Empty

Foot

Full

Gallon

Gram

Heavier

Heaviest

Heavy

Hectogram

Hectoliter

Hectometer

Height

Hour

Inch

Kilogram

Kiloliter

Kilometer

Length

Light

Lighter

Lightest

Long

Longer

Longest

Measure

Meter

Meter stick

Metric

Mile

Millennium

Milligram

Milliliter

Millimeter

Minute

Narrow

Nickel

Nonstandard units

Odometer

Ounce

Pan balance

Peck

Penny

Perimeter

Pint

Platform scale

Pound

Quart

Quarter

Ruler

Scale

Second

Short

Shorter

Shortest

Size

Small

Spring scale

Standard units

Tall

Taller

Tallest

Tape measure

Temperature

Thermometer

Time

Today

Tomorrow

Ton

Trundle wheel

Unit
Volume
Week
Weight
Wide

Yard
Yard stick
Year
Yesterday

Measurement techniques and formulas words:

Approximation
Estimation
Formula
Hypotenuse
Iterating

Length
Pythagorean Theorem
Unitizing
Units
Width

CHAPTER 5

Levels of Cognitive Demand in Measurement

Adjusting LCD in Problem Solving

See:
Number, p. 107
Algebra, p. 50
Data, p. 69
Geometry, p. 72

A group of first graders works with measuring distance. Because the group has had some difficulty with the concept, the teacher uses a lower level of cognitive demand (LCD) by modeling a think aloud strategy and using guiding questions to help the students with the problem solving situation.

The teacher announces to the group that they will try an activity called the *straw javelin*. This activity involves throwing straws and measuring distance. The first student steps up to the starting line (a piece of tape on the floor) and throws a straw and the students watch it land. The teacher asks, "How far did the straw travel?" The students point to the straw, but do not suggest any ways to measure the distance. Therefore the teacher models a think aloud strategy, saying "To figure out how far the straw traveled, we could use a measurement tool. I have decimeter sticks (base-ten rods), centimeter cubes (base-ten block units), and a meter stick." The students grasp the measurement opportunity and begin to measure the distance with the decimeter sticks. They use eight sticks placed end to end from the starting line to the closest part of the straw. The teacher asks, "How far did the straw go?" The students answer, "Eight." The teacher asks, "Eight what?" One student responds, "Eight sticks." The teacher lines the meter stick up next to the decimeter sticks and says, "Yes, there are eight sticks. Is there another way we can describe the distance?" The students look at the meter stick and announce, "Eight decimeters."

One of the other students says, "It is also eighty centimeters. Let's use the cubes next to the sticks." With the teacher's help, the students line up eighty centimeter cubes. The teacher asks, "Is eighty centimeters the same as eight decimeters?" The students say, "Yes." The teacher writes *eight decimeters* on one sticky note and *eighty centimeters* on another sticky note. The students place these sticky notes next to the straw on the floor.

The teacher asks, "What if the straw traveled one more decimeter?" One student replies, "That would be eighty-one decimeters." Another student says, "No. Not eighty-one decimeters. It would be nine decimeters." The teacher asks her to show her thinking with the measurement tools. She lines up another decimeter stick and counts the decimeters. The students agree with her thinking. The teacher continues to model problem solving strategies and pose similar questions as the rest of the students take a turn with the straw javelin.

In this scenario, the teacher did not require an extremely low LCD because the students were not given the answers. Additionally, the teacher uses several different levels of questioning, including *what if* questions. Yet the LCD was not extremely high because the students were given modeling and related questioning. The teacher used the LCD that best fit the needs of the students.

Given the same topic (measuring distance of the straw javelin) but a different group of students, the teacher may differentiate the problem solving situation in another way. This group of students understands the basics of measuring with centimeters and decimeters and is ready to work on problem solving in a way that uses a much higher LCD.

After a few straw tosses, the teacher presents this problem solving situation to the students.

James throws a straw 2 decimeters further than Monica.
Airika throws a straw 4 centimeters less than James.
Monica throws a straw 27 centimeters less than 1 meter.

The students turn to the teacher and ask, "How do we solve this?" The teacher answers, "Figure out what you know and what you need to know." The students decide that from the given information, they know that Monica throws a straw 27 centimeters less than one meter. The students decide to line up 10 decimeter sticks next to a meter stick to figure out the distance. They subtract 2 decimeters because they understand that 2 decimeters is the same as 20 centimeters. "We still need to subtract 7 centimeters," says one of the students. "How?" asks another student. A student suggests that they replace 1 decimeter with 10 centimeters. "That's a great idea," responds one of the students,

"Then we can subtract the seven." The students do so and figure out that Monica tossed the straw 73 centimeters. The students are pleased to have solved part of the problem.

The teacher asks, "How can you use this information to figure out how far James and Airika tossed their straws?" The students decide that they cannot figure out how far Airika threw her straw until they know how far James threw his straw. One of the students says, "All we have to do is add 2 decimeters to Monica's distance." The other students agree. "It is going to be 93 centimeters." Other students agree. One student says, "Let's prove it with the sticks," as he adds 2 decimeter sticks. "The last part is easy," says one of the students. "Why do you think this part is easy?" asks the teacher. The student replies, "Because you just subtract 4 centimeters." The other students agree. "Wait," says one of the students, "We don't have enough centimeters." Another student suggests exchanging a decimeter stick for 10 centimeter cubes and the group agrees.

After the students finish solving the problem, the teacher invites the group to work with partners to create a similar problem solving situation using the straw javelin scenario. Even though there was one correct answer in this problem solving situation, there was a degree of open-endedness because there were several ways to solve the problem. Notice how the teacher did not model how to solve the problem, yet constantly raised the LCD by asking questions that helped the group continue to probe deeper into the math situation. Additionally, the level of ownership moves up a notch when the students ask questions and exchange ideas with each other, instead of directing discourse and questions solely to the teacher.

The levels of cognitive demand required in math problem solving situations should vary based upon the needs of the students. Adjusting the LCD based on the immediate academic needs of the students is an excellent way to differentiate mathematics instruction.

Adjusting LCD in Reasoning and Proof

See:
Number, p. 110
Algebra, p. 52
Data, p. 73
Geometry, p. 76

The teacher presents this time measurement problem:

It takes Max ninety minutes to finish his homework, eat dinner, and read a book.

It takes Casey two hours to finish her homework, eat dinner, and read a book.

Who takes longer to finish? How do you know?

The teacher works with a group of students using reasoning and proof to solve the elapsed time problem. Notice how the teacher starts with a lower LCD and adjusts the LCD to a higher level.

Teacher: Who takes longer to finish?

Denisha: I think it is Max.

Teacher: Tell us about your reasoning.

Denisha: Well, ninety is a lot bigger than two.

Caitlin: It is really a lot bigger.

Teacher: Are the units the same?

Denisha: Oops. I was thinking two minutes.

Caitlin: I was thinking that, too.

Teacher: How will we work with minutes and hours?

Frank: I know there are sixty minutes in one hour.

Teacher: How can we use that information to prove who took longer?

Alexa: We could show it on a clock.

Denisha: Yeah. I will get the clocks.

Teacher: What if we have a clock for Max and a clock for Casey?

Students: Yes.

Frank: I will show Casey's time. What time should we start with?

Caitlin: Let's start with 5:00 because that's when I get home.

Denisha: I will show Max's time.

[The students show sixty minutes of elapsed time on both clocks.]

Alexa: I think Max will be done at 6:30.

Teacher: Tell us what you did to figure that out.

Alexa: We already showed sixty minutes, so if we show thirty more minutes that will be 6:30.

Teacher: Why did you decide to solve the problem that way?

Alexa: Because sixty minutes is an hour and thirty minutes is ½ hour.

Caitlin: Another way to prove it is 60 + 30 = 90.

Denisha: I can prove it using the clock. [She counts by fives to thirty moving the minute hand on the clock.]

Frank: Casey will be done at 7:00.

Caitlin: Casey takes longer.

Alexa: How can we prove it?

Caitlin: Because you have to go another halfway around the clock for Casey.

Teacher: Why do we need to go another halfway around the clock?

Denisha: Because another halfway brings it to 7:00.

Caitlin: That will be 2 hours.

Frank: Two hours is 120 minutes. I know because 60 plus 60 equals 120.

Denisha: Yes! We proved it because 120 minutes is longer than 90 minutes.

In this scenario the teacher prompts the students to use more reasoning and proof as the differentiated small group lesson continues. Based on the students' responses and levels of thinking, the teacher moves the instruction from a lower LCD to a higher LCD based on the immediate learning needs of the students. The teacher also encourages the students to use explanation and justification.

Adjusting LCD in Communication

See:
Number, p. 112
Algebra, p. 53
Data, p. 74
Geometry, p. 76

Too often young children are required to communicate something that they do not yet understand. For example, let's examine a specific time of day—say one o'clock. It may seem like a relatively simple concept. It is an on-the-hour time. We do not have any minutes after the hour to deal with. Yet when we write *1:00,* it looks similar to *100* because there are two zeros. As a first grader once told me, "It looks like one hundred o'clock." Communicating on-the-hour times is complicated because the zeros serve as place holders for the minutes when we write the time. For some children announcing the time as one o'clock and zero minutes can be helpful as they learn how to communicate time.

Communication provides another way to adjust the LCD as we differentiate instruction in mathematics. Of course, we want all children to communicate in mathematics, but we can use the various types and levels of sophistication to adjust the levels of cognitive demand.

The teacher works with a group of students on measuring temperature. The students work in dyads to measure the change in temperature of a cup of water. The students communicate the temperature by describing what happens to the red fluid when warm water is added to the cups of water. The students communicate with phrases such as "The red line goes up," "The red rises fast," and "The red line gets taller." When ice water is added to the cups of water, the students communicate the changes in temperature with phrases such as "The red line goes down," "The red line gets shorter," and "The red line starts to disappear."

The teacher encourages the students to use pictures to describe these and other changes in temperature. The students draw thermometers with various heights of the red line. They label these pictures with the words *cold*, *warm*, and *hot*. The teacher encourages the students to talk about the changes in temperature as they share the pictures with each other. The students use written and spoken words as well as pictures to communicate the changes in temperature. The LCD in communication corresponds to the needs of the students in the group.

With a different group of students, the teacher increases the LCD by including the actual temperatures. These students participate in the same experiences of measuring the changes in the temperature of water. However, because these students need a higher LCD, they communicate the changes in temperature by stating the actual temperatures in degrees Celsius and degrees Fahrenheit. The level of communication is further increased as they compare the temperatures according to the differences. For example, the students find that the water is 72°F. When the warm water is added, the temperature rises to 94°F. The students engage in discourse about the change in temperature with comments such as, "It is now 94°F and it was 72°F, so to figure out the difference, you have to subtract and 94 − 72 = 22." To propel the LCD further, the teacher invites the students to ask each other questions such as, "What if the water dropped forty degrees Fahrenheit?" "What if the water was twice as hot?" and "How are degrees Fahrenheit related to degrees Celsius?"

The LCD in communication is much higher for this group because that is the level of instruction that they currently need. They are able to use sophisticated ways to communicate measurement through words, numbers, and even

See:
Number, p. 117
Algebra, p. 54
Data, p. 75
Geometry, p. 81

equations. The teacher encourages the communication to include student-to-student questions to further increase the LCD.

Adjusting LCD in Connections

Measurement is connected to many topics outside of mathematics such as science, art, literacy, social studies, physical education, music, and technology. Helping students make these connections enhances their understanding and application of measurement. Measurement is also connected to all of the other math content standards in multiple ways. The ties between number, operations, and measurement are especially profound. Counting and finding the totals and differences in various lengths, for example, are measurement concepts that are completely embedded in numbers, number relationships, and number operations. Likewise, measurement and geometry are interconnected. Many shapes are defined by the measurements of the angles in those shapes. Other connections within the standards may appear less direct until one takes a closer look. Data analysis, for example, may seem like a very different field of study when compared to measurement. Yet many data displays include measurement. Algebra is also connected to measurement. For example, many geometric patterns are based on consistent (proportional) changes in measurement. Helping students understand these and other connections provides ways to adjust the LCD when differentiating instruction in mathematics.

In a nutshell, if the connections are simple, the LCD is low. If the connections are complex, the LCD is high. Teachers need to adjust the levels based on the specific needs of the students.

A class is currently working on measurement. The teacher has preassessed the students and learned that some need a higher LCD and others need a lower LCD. Yet the teacher plans for all of the students to grow in their understanding of measurement and making connections between measurement and literacy.

The students will work in small groups with their teacher, Ms. Kingsley. While some students are working with the teacher, other students will work on writing in their math journals. The class is given this prompt: *How is measurement used in stories?* All students will address this question as they write in their journals. The idea will also be addressed in the small group work with the teacher.

Ms. Kingsley meets with the group that needs the lower LCD first. She wants to provide some prewriting experiences to help the students be successful. She opens the small group mini-lesson by asking, "Do you know any stories that include measurement?" The children have a difficult time thinking of a story. Ms. Kingsley asks, "What about the story about the girl with the golden locks of hair?" The students are excited to think about the familiar story. One of the students asks, "Like the chair is too small so she broke it?" "Yes. Exactly," replies Ms. Kingsley. Another student asks, "What about the porridge?" Another student asks, "Was some in a big bowl and some in a small bowl?" "No," answers another student, "Poppa's porridge was too hot." Ms. Kingsley asks, "Is temperature something you can measure?" "Yes!" respond the students, pleased with the connections they have discovered. Ms. Kingsley asks the students to write and draw about these connections in their math journals. She assigns each of the students a partner that they can talk to about the connections between measurement and this particular story as they write in their math journals.

Ms. Kingsley asks the next group to join her. The preassessment indicated that these students need a medium LCD. Ms. Kingsley draws comparison circles on the chalkboard (see Figure 5–1) and labels one circle *Length* and the other circle *Weight*. She asks the students to think about any stories that involve length or weight. "We read a story about a tree that grows," responds a student. The other students make the connection. One student says, "Yes. The tree gets taller. We should put that story in the *length* hoop." Ms. Kingsley writes the title of the story on an index card and tapes it in the *length* hoop.

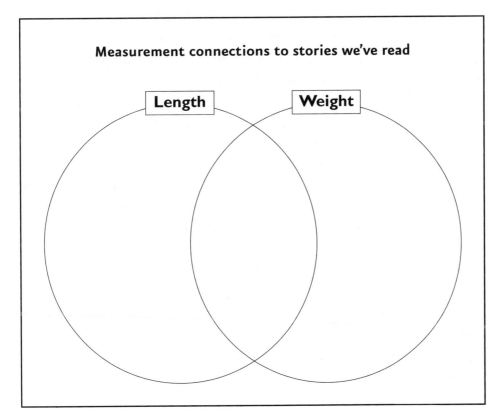

Figure 5–1

A Set of Comparison Circles Regarding Measurement

Measurement connections to stories we've read

Length   **Weight**

Another student says "What about the story of presents? Each present weighed a different amount." The other students recall the story. Ms. Kingsley asks, "Which hoop should we put that story in?" The students tell her to put the title of that story in the *weight* hoop. One of the students says, "Remember the story about the fish? It keeps growing bigger because the boy keeps feeding it." The other students make the connection. One student says, "In that story, the fish gets longer and it gets heavier." "So that story goes in both hoops" says another student. The rest of the students agree. Ms. Kingsley invites the students to draw this graphic organizer in their math journals as a way to show connections between measurement and literacy. She encourages the students to work together to think of other stories and where to place those stories in the graphic organizer.

Ms. Kingsley calls the third group to work with her. The preassessment indicated that these students need a high LCD. Ms. Kingsley draws a triple hoop graphic organizer on the chalkboard. She asks the students to name some fiction and nonfiction books that they have read. Ms. Kingsley writes the titles of nonfiction books on yellow index cards and the titles of fiction books on green index cards. The students come up with ten titles of books. Ms. Kingsley spreads the index cards out on the table and asks, "What type of measurement do we find in each of these books?" The students begin to name length, weight, and capacity as the types of measurement found in the books. Ms. Kingsley uses these as the category labels for the three hoops, as shown in Figure 5–2. She asks the students to place the index cards on the graphic organizer according to the type of measurement found in the books. Some books only have one type of measurement. Other books have two types or all three types of measurement. The students make the connections as they discuss the measurement found in each book. At one point in the discussion, the students decide they need to refer to the book to check the type of measurement. This prompts them to check each of the books to make sure they have found all of the connections. Ms. Kingsley helps the students divide up the books so each student can look for additional connections and evaluate the connections the group has already found. The LCD is extremely high for this group because of the multiple connections and the complexity of the graphic organizer.

Later in the day, two of the students approach Ms. Kingsley and share, "We found another book that connects to measurement. Can we write it on an index card and add it to the hoops?" The LCD has skyrocketed because the students are continuing and broadening the task without any guidance from the teacher. "When students seek new learning experiences in the absence of teacher direction, that is the ultimate transfer agency" (Kramer 2005). Differentiated instruction prompts powerful learning.

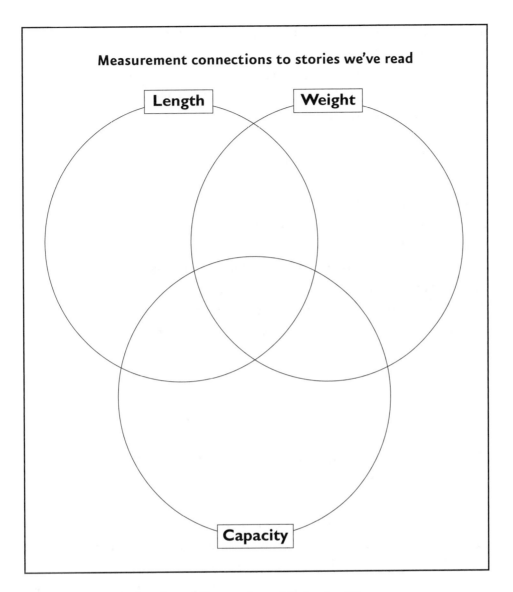

Measurement connections to stories we've read

Length

Weight

Capacity

Figure 5–2 *Another Set of Comparison Circles for Measurement Connections*

Adjusting LCD in Representations

A first grade class is studying time. The teacher wants to differentiate instruction using LCD in representations. She decides that one group of students will make a timeline of the events that occurred during the week. She meets with the group and asks them to share some of the things they have done in school during the week. The students share several things including going to art, music, PE (physical education), computer lab, media center, and science. They also include the special events from the week such as the bus drill and the "Imagination" assembly. The teacher helps the students construct the timeline

See:
Number, p. 119
Algebra, p. 56
Data, p. 78
Geometry, p. 83

framework. She asks, "Which day should we put first on the timeline?" The students respond, "Monday." The teacher asks, "Why should we put Monday first?" The students explain that even though Sunday is the first day of the week, it is not a school day, so they will start with Monday on the timeline. The teacher asks questions and directs the students as they develop the timeline framework.

The other group of students makes a timeline too. Their timeline is of the day's events. When the teacher meets with these students she says, "You will make a timeline of our school day. Think about the important events of the day and the times that these things happen." The teacher gives the group a set of time cards and asks the group to pick out the time cards that they want to use in their timeline. The students begin working on the task. They order the time cards and discuss which events are the most important during the day. The students sketch some of the events they believe are the most important. They discover that their sketches are not as clear as they had hoped. They decide to ask the teacher if they can use one of the school's digital cameras to take pictures of the events of the school day. The teacher explains that they can use the camera if they take turns and use the camera carefully. The students make a list of the times that they want to take a picture and who will take the picture. The teacher leaves the camera on her desk and allows the students to pick up the camera and take photographs in accordance with the list.

At the end of the day, the teacher types up the events the students making the timeline of the week listed and cuts them into separate sentence strips. She also prints out the photographs taken by the group making the timeline of the day. On Monday, the teacher gives the photographs to the group that worked on the day's timeline and asks them to finish constructing the timeline. They decide that they missed two important events (pledge of allegiance and lunchtime) and ask to borrow the camera again.

The teacher works with the group that makes the timeline of the week using the sentence strips. The students are not sure where to place the *We go to music* and the *We go to PE* sentence strips since they go to music and PE on Mondays and Thursdays. The teacher suggests making an extra copy of these sentence strips. The students are pleased with the idea.

Both groups of students worked with measuring time and constructed the timeline in Figure 5–3. Yet each group worked on the task at their specific, appropriate LCD in representations. One group received more guidance, direction, and support. The other group received some direction in the task, but the representation was more student-generated and student-directed. The teacher successfully differentiated the instruction according to LCD in representations.

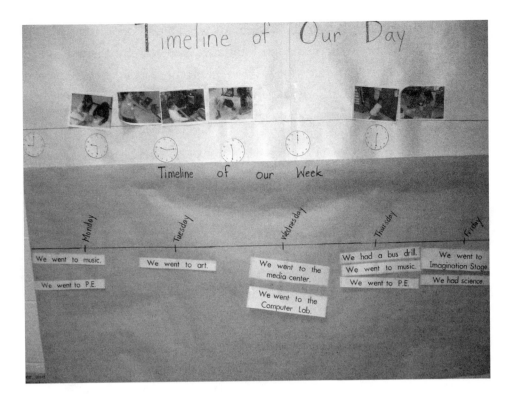

Figure 5–3 *The Timeline That the Students Constructed*

Using the Levels to Differentiate Math Tasks

Let's look at one measurement concept, *capacity*, and see how it can be adjusted to the various levels of cognitive demand.

See:
Number, p. 125
Algebra, p. 58
Data, p. 79
Geometry, p. 84

    Level 1 task: Define capacity.

    Level 2 task: Explain what capacity means.

    Level 3 task: Show how to find the capacity of a specific container.

    Level 4 task: Classify containers according to capacity.

    Level 5 task: Evaluate someone's way of finding capacity.

    Level 6 task: Create a new unit for measuring capacity.

Using these tasks to differentiate instruction does not mean that all students will proceed through the tasks in consecutive order. Some groups of students may work on the first couple of levels. Other students may focus on the middle levels. Still other students may only work on the higher-level tasks. The important thing is that the children are given tasks that are appropriate for their particular level.

The levels of cognitive demand offer many windows of opportunity to differentiate instruction in mathematics. I am always in search of new ways to *bump up* the LCD or *tone down* the LCD to enable my students to better grasp the math that I am teaching. Ultimately, my goal is for all students to receive just the right amount of support and challenge as they learn and grow.

CHAPTER 6

Learning Frameworks in Measurement

Incorporating Learning Styles

A first grade class is currently working on a unit on measuring length. The students are at various levels of understanding measurement of length. Some are currently working with nonstandard units and some are currently working with standard units. The teacher has assessment data that indicate the students' learning styles. The groups are heterogeneous by academic ability and homogeneous by learning style.

See:
Number, p. 128
Algebra, p. 59
Data, p. 80
Geometry, p. 85

Visual Learners Work with Measuring Length

When the teacher meets with the students who are visual learners, she gives each student a piece of string. She explains to the students that the length of each string is an approximation called one *cubit*. The students ask, "What's a cubit?" The teacher responds, "That's a great question. In fact, that is what you are going to try to figure out during our group time." The teacher gives the students a set of picture cards that represent specific lengths of objects—such as a crayon, a chair, a laptop computer, a building, and body measurements such as finger widths, palm widths, hand spans, elbow to fingertip (unbeknownst to the students this is the cubit), outstretched arms, and a thumb. The students are asked to look through the picture cards and create categories of which ones could be a cubit and which ones could not be a cubit. The teacher explains that the students need to visualize the measurements shown on the picture cards because the measurements are not to scale. As the students work on the task, they discuss what the measurements look like. They look at the length of the string and decide that there are two picture cards that could represent a cubit. The cards they chose are the laptop computer and the elbow to fingertip. The students explain their choices and ask for an additional clue. The teacher explains that a cubit is a measurement that has been used for hundreds of years. The students decide that laptop computers have not been around that long, so they deduce that a cubit is a measurement from the elbow to fingertip. The teacher invites the students to draw a graphic organizer that includes items that are bigger than a cubit and items that are smaller than a cubit.

In this small group lesson, the students are able to use their visual preferences to help them learn more about measurement of length.

Auditory Learners Work with Measuring Length

When the teacher meets with the auditory learners the focus is on sound, pitch, and voice. The teacher tells the students a story.

> A long, long time ago there lived a little boy and his mother. The boy wanted to help his mother measure and cut pieces of string to make sachet pouches. The problem was there were no rulers or other measurement tools available. The mother explained to the boy that she needed three sizes of string—small, medium, and large. The small size she needed was an approximation called a finger width. The medium size that she needed was an approximation called a palm width. The large size that she needed was an approximation called a hand width. The boy tried and tried, but he could not remember which lengths to cut the string.

The teacher invites the students to create a poem to help the little boy remember the three sizes. The students begin working on the task. They discuss the lengths and ways the little boy could remember the measurements. With the teacher's help they create a cute poem to help the boy remember specific lengths.

> Hey little boy.
> Your finger is small.
> Check how wide it is.
> Not how tall.
>
> Hey little boy.
> Medium is your palm.
> Use the center of your hand.
> And you won't be wrong.

Hey little boy.
Use your finger to your thumb.
That's the largest size
Now you can help your mom.

Creating the poem helps the students use their auditory strengths to learn more about measuring lengths. As the students recite the poem they use sound, pitch, and tone to connect the three measurements. In this learning experience, the students use their auditory preferences to help them learn more about measuring lengths.

Kinesthetic Learners Work with Measuring Length

The teacher explains to the kinesthetic learners that a long time ago people used body measurements because they did not have measurement tools. She introduces the word *fathom* and explains to the group that a fathom is an approximation of the measurement from fingertip to fingertip when your arms are stretched out. The students stretch their arms to show fathoms. The teacher invites the students to measure one of the walls in the classroom using fathoms as the units. The students begin measuring the wall from side to side. The teacher encourages the students to figure out how they will show the rest of the class how to measure with fathoms.

The students in this group use their kinesthetic preferences to learn more about measuring length.

Tactile Learners Work with Measuring Length

The teacher also explains to the tactile learners that a long time ago people used body measurements as approximations because they did not have measurement tools. With this group she shares that the word *inch* was originally used to name the width of a thumb. The teacher invites the group to use their tactile strengths to measure various objects around the room using thumb widths. They are invited to record these lengths using thumbprints. In this way, the students use hands-on experiences to learn more about measuring length.

In this example, the learning style preferences were addressed in ways that helped the children to better understand measurement. They also had the opportunity to learn more about the history of measurement and the use of approximations. The instruction was differentiated to meet the learning needs of the students. Additionally, each group created a final product that reflected the learning style preference—the visual group created a graphic organizer, the auditory group created a poem, the kinesthetic group developed a demonstration, and the tactile group created thumbprint measurements.

See:
Number, p. 132
Algebra, p. 60
Data, p. 82
Geometry, p. 86

Incorporating Multiple Intelligences

In a second grade class the students are learning about measuring weight. The students have participated in a recent inventory of multiple intelligences and the teacher grouped students by likenesses in intelligences. During the lesson, the teacher serves as a facilitator moving from group to group answering and asking questions, advising, assisting, and directing. The teacher shares with the class that the produce manager at the local grocery store needs to know the weights of the fruits to help him set up a new display. All groups are given the same four pieces of fruit (apple, strawberry, lemon, and cherry) and asked to show (*showing the knowing*) the weight of each piece of fruit in a different way.

Showing the Knowing in the Word Smart Group

Five students show strengths in word smart (verbal-linguistic intelligence). Their task is to write a letter to the produce manager at the grocery store describing the weights of the fruit. Here is the letter the group creates, revises, and eventually reads to the class.

Dear Produce Manager,

You need to set the fruit up in a good way. We think you should put the heaviest fruit on the bottom and the lightest fruit on top. We used a simple balance to compare the weights. The apples are the heaviest. The lemons are next. Then the strawberries come next. The cherries are the lightest, so they should go on top. We hope this information helps you.

Sincerely,

Second graders

Showing the Knowing in the Number Smart Group

Four students are high in number smart (mathematical-logical intelligence). These students use a pan balance and pound weights to find the weight of each piece of fruit. They find that the cherry weighs .02 pounds, the strawberry weighs .1 pounds, the lemon weighs .25 pounds, and the apple weighs .5 pounds. To verify their findings they use a platform scale. All of the numbers align except the lemon. The platform scale says the lemon weighs .249 pounds. The teacher takes the opportunity to teach the students about rounding. Because the students are number smart, they understand the concepts relatively easily. They decide to check the other fruits again to see if rounding is necessary. The group records the findings in a chart to give to the produce manager at the local grocery store.

Showing the Knowing in the Picture Smart Group

There are four students who indicated some strength in picture smart (visual-spatial intelligence). These students were given the task of making pictures of the fruits and constructing a poster. The students draw pictures of the fruits on separate index cards. The pictures are related to what they learn as they use a pan balance to compare the weights. The apple is the heaviest, so they make the picture of the apple large and dark. The picture of the lemon is a little bit smaller than the apple and the coloring is a little bit lighter. The picture of the strawberry is smaller and lighter than the first two pictures. The picture of the cherry is the smallest and lightest of all. The students arrange the pictures according to the weights of the fruits on the poster to give to the produce manager at the local grocery store.

Showing the Knowing in the Body Smart/Music Smart Group

There were three body smart (body-kinesthetic intelligence) students and four music smart (musical-rhythmic intelligence) students who were grouped together and given the task of creating a chant with movements to describe and explain the weights of the fruits.

After weighing the fruits with a pan balance and pound weights, the students developed the following chant. They used a *stomp, stomp, clap* pattern between the verses of the following chant:

We weighed
We weighed
The fruit.
Stomp, stomp, clap. Stomp, stomp, clap.

We weighed
We weighed
The fruit.
Stomp, stomp, clap. Stomp, stomp, clap.

Apple is
Apple is
Half pound.
Stomp, stomp, clap. Stomp, stomp, clap.

Lemon is
Lemon is
Quarter pound.
Stomp, stomp, clap. Stomp, stomp, clap.

Strawberry is
Strawberry is
One tenth pound.
Stomp, stomp, clap. Stomp, stomp, clap.

Cherry is
Cherry is
Two hundredths pound.
Stomp, stomp, clap. Stomp, stomp, clap.

We weighed
We weighed
The fruit.
Stomp, stomp, clap. Stomp, stomp, clap.

We weighed
We weighed
The fruit.
Stomp, stomp, clap. Stomp, stomp, clap.

With the teacher's help, the students incorporated music and large motor skills to show what they know about the weights of the fruits. They plan to perform the chant for the produce manager at the local grocery store.

Showing the Knowing in the People Smart Group

The final group consisted of the people smart (interpersonal intelligence) students. The students were asked to develop a survey to ask other people to estimate the weight of each of the fruits. The teacher allowed the students to go to the school's office to survey the office staff and administration. They asked ten adults to estimate the weight of each fruit. When they returned to the classroom, they used the pan balance and gram weights to learn if the estimates made by the adults were in a close range. The group plans to share these findings with the produce manager of the local grocery store.

In this second grade example, the multiple intelligences were used as a means of *showing the knowing*. The students demonstrated understanding via their individual and combined personal strengths. The instruction was differentiated to meet the learning needs of the students because the tasks and products were matched to specific intelligences.

Incorporating Affective Needs

In a first grade classroom, three students (Justice, David, and Randell) are using the pan balance and weights to weigh a calculator. David continues to add weights after the balance is equal. Randell uses an *I message* to explain his feelings. Randell says, "I feel upset when you keep adding weights because I want to find the measurement." David begins taking some of the weights off the balance and says, "OK. I won't do it again." Randell and Justice reply, "Thanks," and all three boys work together to find the weight of the calculator.

In this way, the *I message* prompted a change in the learning situation based on the affective needs of Randell and Justice. It allowed the mathematics learning to take place, highlighting yet another way to differentiate mathematics instruction in the prekindergarten through second grade classroom.

See:
Number, p. 141
Algebra, p. 64
Data, p. 85
Geometry, p. 89

Incorporating Interests

In a second-grade classroom, the teacher has the students decide which sport they would like to research. Each group will look on the Internet and in books to find out the important measurements associated with specific sports. Students choose basketball, baseball, bowling, football, and tennis. While the groups work together in the computer lab/media center, the teacher moves from group to group facilitating and offering support. The basketball group learns that the dimensions of a college basketball court are 94 feet long and 50 feet wide. The ring of the basket has an inside diameter of 18 inches. The baseball group learns that the distance from each base to the next base is 90 feet. Each base is 15 inches by 15 inches. The bowling group learns that the overall length from the foul line to the pit is 62 feet and $10\frac{3}{16}$ inches. The gutters are $9\frac{5}{16}$ inches wide. The football group learns that the length of the field is 120 yards, including 10 yards in each end zone. The width of the football field is $53\frac{1}{3}$ yards. The students interested in tennis find out that the court is 120 feet long and 60 feet wide. The height of the tennis net is 3 feet in the center and 3 feet 6 inches at each post. The students share their findings with the entire class.

Incorporating students' learning frameworks serves as a powerful mechanism for differentiating instruction in mathematics. Learning styles, multiple intelligences, environmental preferences, affective needs, and interests are ways to connect students with positive, prolific learning experiences. Students' motivation and commitment are increased as teachers incorporate learning frameworks.

See:
Number, p. 147
Algebra, p. 64
Data, p. 86
Geometry, p. 89

CHAPTER 7

Personal Assessment in Measurement

T-Charts Show Personal Assessments of Measurement

See:
Number, p. 148
Algebra, p. 68
Data, p. 87
Geometry, p. 90

One of the most basic graphic organizers is a simple yes/no chart. Using the yes/no labels in the form of a T-chart, students are asked if they are comfortable with a specific math situation. For example, before beginning a review of how to measure weight, the teacher asks the students to indicate if they know how to use a scale to measure. Each student places a name card in the *yes* or *no* column. Sometimes instead of name cards the class can use clothespins with individual students' names on each. Each clothespin is clipped on the *yes* or *no* side of the chart by the students as in Figure 7–1. The teacher can take a quick look and see which students believe they need to review the topic and which students believe they need an additional challenge.

After working with his class on measuring length, the teacher asked the students to place their names on the yes/no chart. The question on the chart, shown in Figure 7–2, was *Can you measure length to the closest ½ inch?*

Many of the students felt comfortable measuring length to the closest ½ inch. The teacher met with the six students who were not yet confident measuring to the closest ½ inch in a small group setting. He first complimented the students on their courage and honesty. Then he targeted the instruction to meet their academic needs. By the end of the small group session, several of the students felt comfortable enough to move their names to the *yes* side of the chart.

Figure 7–1 *A T-Chart Used in a Class to Show Students Topic Knowledge*

Using graphic organizers offers a twofold benefit. The teacher gathers important personal assessment data and students have more opportunities to engage in mathematics because the data can be used instructionally. Many times the teacher talks with the students about what the data show. Using math vocabulary as they analyze and interpret the information revealed in the graph offers children additional opportunities to experience real-world mathematics.

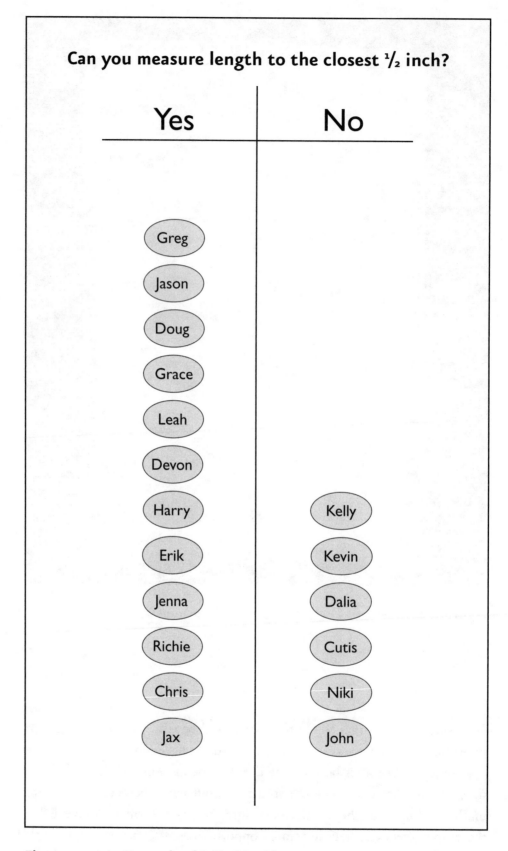

Figure 7–2 *An Example of A Yes/No Chart*

Bar Graphs Show Personal Assessments of Measurement

Bar graphs are also powerful tools. For example, before beginning a lesson on measuring weight the teacher may put up a bar graph shell titled *How much I know about using a balance*. The labels include a large balloon (I know a lot), a medium-size balloon (I know some), and a small balloon (I know a little bit). Students use sticky note cards to place their names in the category that represents how much they know (see Figure 7–3).

The information is used to form groups and target instruction. After the lesson, students can change the placement of their name cards if they have revised their comfort levels with the topic. They may write in their math journals about how their level of understanding of a given topic has changed.

See:
Number, p. 150
Algebra, p. 71
Data, p. 89
Geometry, p. 93

Comparison Circles Show Personal Assessments of Measurement

Comparison circles encourage math thinking and give excellent information. The teacher may use a single hoop or two (or more!) intersecting hoops. Consider this classroom scenario. During a lesson on estimating length, the teacher

See:
Number, p. 152
Algebra, p. 71
Data, p. 89
Geometry, p. 93

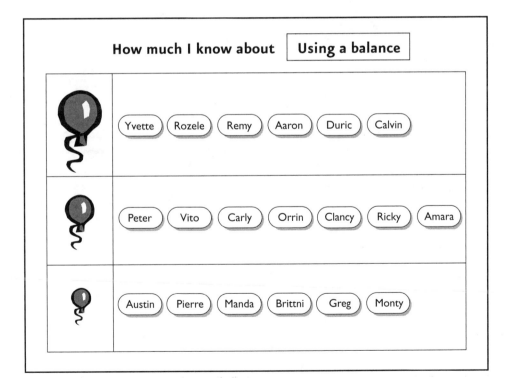

Figure 7–3

An Example of a Bar Graph

asked the students to place their names inside the hoop or outside of the hoop to indicate their current knowledge. The title of the hoop, shown in Figure 7–4, was *I know how to estimate length*.

Using the data presented by the students, the teacher offered a two-tiered lesson. For the students that indicated they knew how to estimate length, the teacher worked with them on estimating perimeter. For the students that indicated they did not know how to estimate length, the teacher worked with them on estimating length. In the two-tiered lesson, everyone worked on estimating measurement and the specific small group lessons were suited to meet the academic needs of the students as perceived by the students.

The teacher used three intersecting hoops as the graphic organizer and asked the students to tell what they know about measuring capacity. The title of the comparison circle graphic (see Figure 7–5) was *I know how to use these capacity measurements*. One hoop was labeled *one cup*, one hoop was labeled ½ *cup*, and the other hoop was labeled ¼ *cup*.

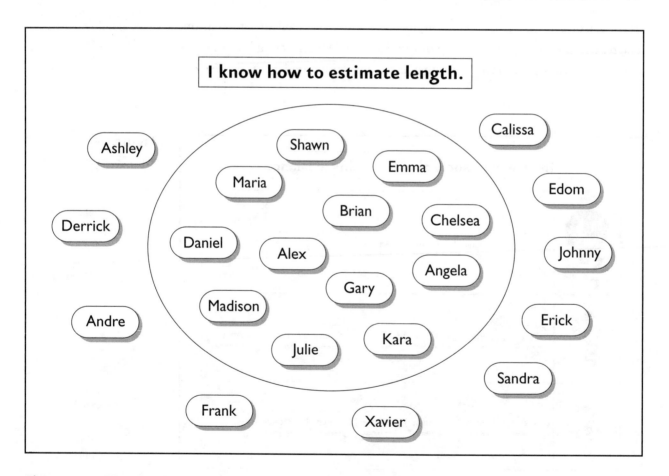

Figure 7–4 *Using a Hoop to Show Students' Current Knowledge*

The data indicate that all of the students are comfortable measuring capacity with one cup. Some of the students can also measure capacity with ½ cup. Some students can measure capacity with one cup, ½ cup, and ¼ cup. The teacher can offer differentiated instruction to the students using the data presented in the comparison circles. The teacher has many options for how to differentiate the instruction. She can meet with each of the small groups and focus on what the students do not yet know. For example, the group that knows how to measure capacity with ½ cup can work on measuring capacity with ¼ cup. The group that already knows how to measure capacity with 1 cup, ½ cup, and ¼ cup can work on more challenging measurements (⅛ cup and ¹⁄₁₆ cup).

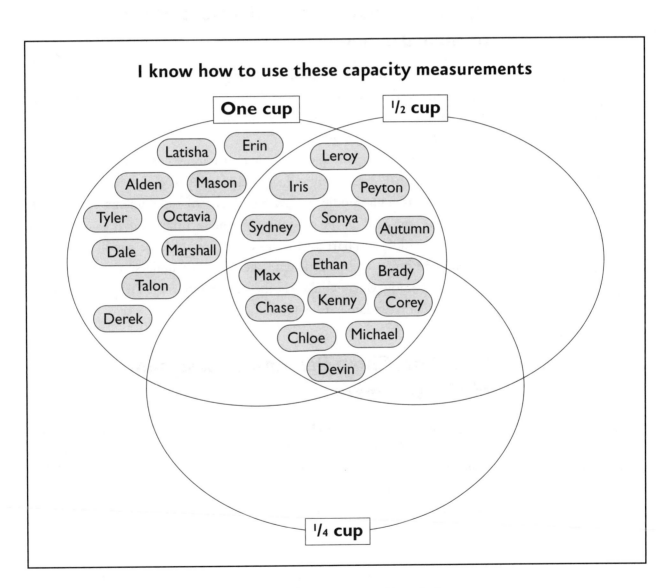

Figure 7–5 *Comparison Circles Show Students' Current Knowledge*

Another option for the teacher is to give students a task they can work on independently or with each other while the teacher meets with small groups. If the students are working without the teacher, it may be most appropriate for the students to expand what they already know rather than move into something they are not yet comfortable with. For example, the students who know how to measure capacity with all three units could work on doubling or tripling the measurements. In this way, the students are working on something new, but they have some comfort with the topic. Therefore they can successfully engage in meaningful tasks without direct instruction from the teacher.

Pyramids Show Personal Assessments of Measurement

See:
Number, p. 154
Algebra, p. 74
Data, p. 91
Geometry, p. 96

Using the pyramid as the graphic organizer, the teacher asked the students to place name cards in ways that show their level of understanding. The title of the pyramid, shown in Figure 7–6, was *Our level of comfort using the trundle wheel to measure length*.

The pyramid graphic allows students to show even more than one of three levels. They can actually place their names to designate levels within levels, such as high-middle or low-high. One of the students strategically placed his name card on the lines dividing the sections. Harris shared, "I am part in the middle and part low." This comment and others indicate that the students are quite able to articulate their level of understanding. It is empowering to know where you are in the learning journey. The teacher can use the data to differentiate mathematics instruction.

Line Plots Show Personal Assessments of Measurement

See:
Number, p. 155
Algebra, p. 75
Data, p. 93
Geometry, p. 97

Using a complex line plot that ranged from zero to one, the teacher asked the students to indicate their level of knowledge. The title of the line plot was *How much we know about measuring temperature with thermometers*. While the students had many experiences working with number lines and values between zero and one, this was one of the first personal assessment opportunities given to this class, so the teacher decided to make the pieces of data anonymous. The teacher hung the line plot behind the easel and allowed each student to privately place an X on the line, as shown in Figure 7–7. Sometimes students need several opportunities to use a graphic organizer in an anonymous

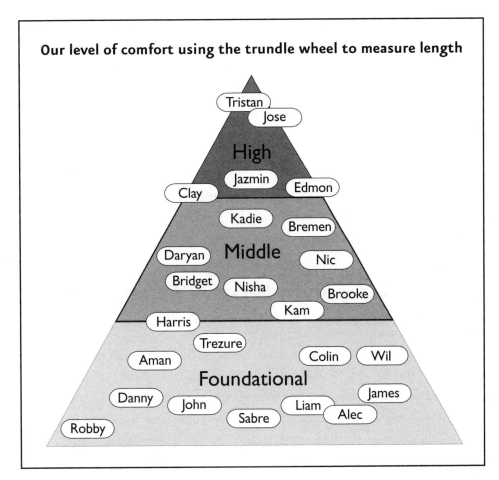

Figure 7–6 *An Example of a Pyramid*

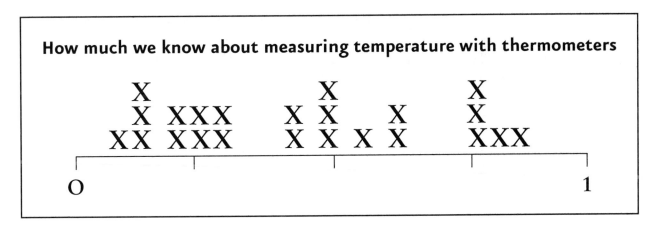

Figure 7–7 *An Example of a Line Plot for Anonymous Reporting*

See:
Number, p. 157
Algebra, p. 76
Data, p. 95
Geometry, p. 98

fashion before they are ready to publicly display their levels of knowledge or comfort.

The data show three distinct groups. The teacher can use the data to offer three different levels of the same task and allow students to choose the level that matches their comfort. The teacher can help students form heterogeneous groups to work on a given task. The teacher can offer small group work at various levels and allow the children to choose which group to attend. Or perhaps the teacher uses an informal assessment to differentiate the instruction and the children use a different-color X to privately indicate their new level of knowledge following the instruction. Afterward, the class can analyze and interpret the data, highlighting growth trends.

Fractions Grids Show Personal Assessments of Measurement

After a lesson on calendars, the teacher posed the fractions grid titled *We know about days, weeks, and months*. Students who were completely confident with days, weeks, and months placed their names in the one whole section of the graphic. Students with solid knowledge (but not complete confidence) of days, weeks, and months placed their names in the ¾ section of the graphic. Students with some knowledge of days, weeks, and months placed their names in the ½ section of the graphic. Students with a little bit of knowledge of days, weeks, and months placed their names in the ¼ section of the graphic. If any students had zero knowledge of days, weeks, and months, they could place their names outside of the graphic (see Figure 7–8).

The teacher used the data to plan for the next day's instruction. She offered a challenge for the students who indicated that they completely understood days, weeks, and months. She offered targeted instruction to the ¾ group in a way that helped them move to complete understanding. She built more foundations with the ½ group and the ¼ group by targeting instruction. All of the students increased their levels of understanding.

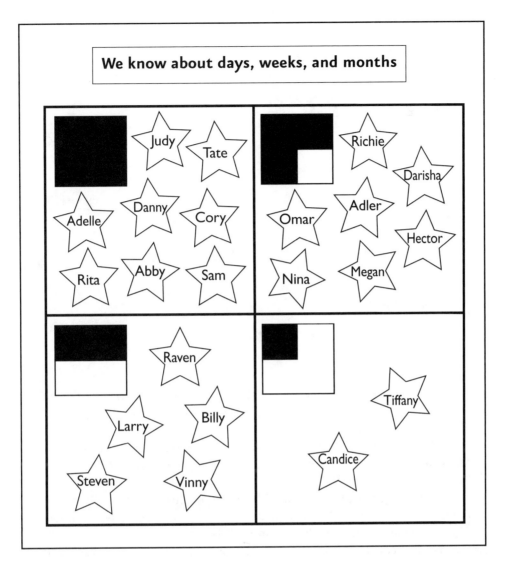

Figure 7–8 *An Example of a Fractions Grid*

REFERENCES

Anderson, L., and D. Krathwohl, eds. 2001. *A Taxonomy for Learning, Teaching, and Assessing: A Revision of Bloom's Taxonomy of Educational Objectives.* New York: Longman.

Barnett-Clarke, C., and A. Ramirez. 2004. "Language Pitfalls and Pathways to Mathematics." In *Perspectives on the Teaching of Mathematics*, edited by R. Rubenstein. Reston, VA: National Council of Teachers of Mathematics.

Bloom, B., M. Englehart, E. Furst, W. Hill, and D. Krathwohl. 1956. *Taxonomy of Educational Objectives: The Classification of Educational Goals—Handbook I: Cognitive Domain.* New York: Longmans Green.

Brewer, D., D. Rees, and L. Argys. 1995. "Detracking America's Schools: The Reform Without Costs?" *Phi Delta Kappa* 77 (3): 210–215.

Caine, R., and G. Caine. 1991. *Making Connections: Teaching and the Human Brain.* Alexandria, VA: Association for Supervision and Curriculum Development.

Copley, J. 2000. *The Young Child and Mathematics.* Washington, D.C.: National Association for the Education of Young Children.

Dienes, Z. 1960. *Building Up Mathematics.* London: Hutchinson Educational.

Dienes, Z., and E. Goldring. 1971. "Learning and Proficiency in Mathematics." *The Mathematics Teacher* 51 (8): 620–626.

Dunn, R., K. Dunn, and G. Price. 1985. *Learning Style Inventory.* Lawrence, KS: Price Systems.

Dunn, R., and S. A. Griggs, eds. 2000. *Practical Approaches to Using Learning Styles in Higher Education.* Westport, CT: Bergin & Garvey.

Gallagher, J. 1993. "Ability Grouping: A Tool for Educational Success." *College Board Review* 168 (Summer): 21–27.

Gardner, H. 1983. *Frames of Mind.* New York: Basic Books.

Gardner, H. 1993. *Multiple Intelligences: The Theory in Practice.* New York: Basic Books.

Gardner, H. 1999. *Intelligence Reframed: Multiple Intelligences for the 21st Century*. New York: Basic Books.

Gentry, M., and S. Owen. 1999. "An Investigation of the Effects of Total School Flexible Cluster Grouping on Identification, Achievement, and Classroom Practices." *Gifted Child Quarterly* 43 (4): 224–242.

Goldring, E. 1990. "Assessing the Status of Information on Classroom Organizational Frameworks for Gifted Students." *Journal of Educational Research* 83 (6): 313–326.

Gordon, T. 1974. *T.E.T.: Teacher Effectiveness Training*. New York: Peter H. Wyden.

Hart, L. 1981. "Do Not Teach Them, Help Them Learn." *Learning* 9 (8): 39–40.

Jarvis, P. 1992. "Reflective Practice and Nursing." *Nurse Education Today* 12 (3): 174–181.

Jensen, E. 1998. *Teaching with the Brain in Mind*. Alexandria, VA: Association for Supervision and Curriculum Development.

Kagan, S. 1997. *Cooperative Learning*. San Clemente, CA: Resources for Teachers.

Kramer, M. J. 2005. "The Key to Raising Achievement: Four Guiding Principles." *AASA New Superintendents E-Journal* (3). http://www.aasa.org/publications/content.cfm?ItemNumber=7039, April 25, 2008.

Kulik, J. A. 1992. *An Analysis of the Research on Ability Grouping: Historical and Contemporary Perspectives*. Storrs, CT: National Research Center on the Gifted and Talented.

Kulik, J. A., and C. C. Kulik. 1992. "Meta-Analytic Findings on Grouping Programs." *Gifted Child Quarterly* 36 (2): 73–77.

Lou, Y., P. C. Abrami, J. C. Spence, C. Poulsen, B. Chambers, and S. d'Apollonia. 1996. "Within-Class Grouping: A Meta-Analysis." *Review of Educational Research* 66 (4): 423–458.

Loveless, T. 1998. *The Tracking and Ability Grouping Debate*. Washington, D.C.: Thomas B. Fordham Foundation.

Loveless, T. 1999. "Will Tracking Reform Promote Social Equity?" *Educational Leadership* 56 (7): 28–32.

National Association for the Education of Young Children and National Council of Teachers of Mathematics. 2002. *Early Childhood Mathematics: Promoting Good Beginnings, A Joint Position Statement*. Reston, VA: National Council of Teachers of Mathematics.

National Council of Teachers of Mathematics. 2000. *Principles and Standards for School Mathematics*. Reston, VA: National Council of Teachers of Mathematics.

Oakes, J. 1985. *Keeping Track: How Schools Structure Inequality*. New Haven, CT: Yale University Press.

Oakes, J. 1988. "Tracking in Mathematics and Science Education: A Structural Contribution to Unequal Schooling." In *Class, Race, and Gender in American Education*, edited by L. Weiss. Albany, NY: State University of New York Press.

Oakes, J. 1990. *Multiplying Inequalities: The Effects of Race, Social Class, and Tracking on Opportunities to Learn Mathematics and Sciences*. Santa Monica, CA: Rand Corporation.

Ogle, D. S. 1986. "K-W-L Group Instructional Strategy." In *Teaching Reading as Thinking*, edited by A. S. Palincsar, D. S. Ogle, B. F. Jones, and E. G. Carr. Alexandria, VA: Association for Supervision and Curriculum Development.

Olszewski-Kubilius, P. 2003. "Is Your School Using Best Practices of Instruction for Gifted Students?" *Talent Newsletter*. Evanston, IL: Center for Talent Development, Northwestern University. http://www.ctd.northwestern.edu/resources/talentdevelopment/bestinstruction.html, May 5, 2008.

Piaget, J. 1977. *The Essential Piaget*, edited by Howard E. Gruber and J. Jacques Voneche Gruber. New York: Basic Books.

Piaget, J. 1983. "Piaget's Theory." In *Handbook of Child Psychology*, 4th ed., edited by P. Mussen. New York: Wiley.

Reis, S. M., D. E. Burns, and J. S. Renzulli. 1992. *Curriculum Compacting: The Complete Guide to Modifying the Regular Curriculum for High Ability Students*. Mansfield Center, CT: Creative Learning Press.

Renninger, K., and S. Hidi. 1992. *The Role of Interest in Learning and Development*. Mahwah, NJ: Lawrence Erlbaum Associates.

Renzulli, J. S., and L. H. Smith. 1978. *The Compactor*. Mansfield Center, CT: Creative Learning Press.

Renzulli, J. S., L. H. Smith, and S. M. Reis. 1982. "Curriculum Compacting: An Essential Strategy for Working with Gifted Students." *The Elementary School Journal* 82: 185–194.

Richardson, K. 2004. "Making Sense." In *Engaging Young Children in Mathematics: Standards for Early Childhood Mathematics Education*, edited by D. Clements and J. Sarama. Mahwah, NJ: Lawrence Erlbaum Associates.

Slavin, R. E. 1987a. "Ability Grouping and Its Alternatives: Must We Track?" *American Educator* 11 (2): 32–36, 47–48.

Slavin, R. E. 1987b. "Ability Grouping and Student Achievement in Elementary Schools: A Best Evidence Synthesis." *Review of Educational Research* 57 (3): 293–336.

Slavin, R., H. Braddock, and H. Jomills. 1993. "Ability Grouping: On the Wrong Track." *College Board Review* 168 (Summer): 11–18.

Smith-Maddox, R., and A. Wheelock. 1995. "Untracking and Students' Futures: Closing the Gap Between Aspirations and Expectations." *Phi Delta Kappa* 77 (3): 222–228.

Taylor-Cox, J. 2001. "How Many Marbles in the Jar? Estimation in the Early Grades." *Teaching Children Mathematics* 8 (4): 208–214.

Tomlinson, C. 1996. *How to Differentiate Instruction in Mixed-Ability Classrooms Professional Inquiry Kit*. Alexandria, VA: Association for Supervision and Curriculum Development.

Tomlinson, C. 1999. *The Differentiated Classroom: Responding to the Needs of All Learners*. Alexandria, VA: Association for Supervision and Curriculum Development.

Tomlinson, C., and J. McTighe. 2006. *Integrating Differentiated Instruction and Understanding by Design: Connecting Content and Kids*. Alexandria, VA: Association for Supervision and Curriculum Development.

Vygotsky, L. 1962. *Thought and Language*. Cambridge, MA: MIT Press.

Wheelock, A. 1992. *Crossing the Tracks: How Untracking Can Save America's Schools*. New York: New Press.

Wiggins, G. 2003. "'Get Real!' Assessing for Quantitative Literacy." In *Quantitative Literacy: Why Numeracy Matters for Schools and Colleges*, edited by B. Madison and L. Arthur. Princeton, NJ: National Council on Education and the Disciplines.

Wiggins, G., and J. McTighe. 1998. *Understanding by Design.* Alexandria, VA: Association for Supervision and Curriculum Development.

Wood, D., J. Bruner, and G. Ross. 1976. "The Role of Tutoring in Problem-Solving." *Journal of Child Psychology and Psychiatry* 17 (2): 89–100.

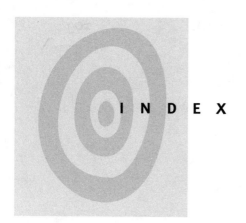

INDEX